The Handbook of ORGANIC HUSBANDRY

The Handbook of ORGANIC HUSBANDRY

FRANCIS BLAKE

The Crowood Press

First published in 1987 by
The Crowood Press
Ramsbury, Marlborough,
Wiltshire SN8 2HE

British Library Cataloguing in Publication Data

Blake, Francis
The handbook of organic husbandry.
1. Organic farming 2. Crops
I. Title
631.5'84 S605.5

ISBN 1-85223-000-2

Dedicated to Charlie

Line illustrations by Vanetta Joffe

Typeset by Acorn Bookwork, Salisbury, Wiltshire
Printed in Great Britain

Contents

Acknowledgements

To write a book on organic farming is a daunting task indeed. The first half of this century saw the publication of a number of classics on the subject. These set a hard precedent to follow, and virtually nothing has appeared ever since.

I would like to thank my wife Jane for initially coming up with the idea of writing such a book, and for persuading me that I had the qualifications and the experience to do it. I must then thank my publishers for making it possible and for the confidence that they seemed happy to place in me.

Organic farming is a complex and diverse subject and I drew on many sources to come up with what I hope is a reasonably comprehensive picture. I valued the advice and help of many experienced organic growers and farmers, either personally or from their own writings. They are too numerous to mention, but I thank them all.

Next, I wish to thank my own farm, the setting up, development and running of which has taught me most of what I know. The experience that I gained in doing this largely forms the basis of this book. I also thank my brother John, who now continues to run it, for the work and responsibility that we shared during so much of that time, which made it all easier and more enjoyable.

One organisation deserves special mention: the Soil Association, which has pioneered organic farming from the outset, and has been in the forefront of the current upsurge in public interest. I have gained much in recent years from serving on its Council, not least in writing skills as part of its editorial group.

Lastly I wish particularly to thank Charles Wacher, secretary of the Organic Growers' Association, for reading through the manuscript with such a critical and constructive eye. But for a crippling disease, which has since claimed his life (though never his spirit) it is he who would have written this book.

Foreword

Two of E. F. Schumacher's best epigrams came to my mind as I started to write this foreword: 'We speak of the battle with Nature', he observes in *Small is Beautiful*, 'but we would do well to remember that if we win that battle, we are on the losing side'. The warning is as appropriate now as it was then, fifteen years ago, and nowhere does it apply with greater force than to agriculture and food production in modern industrial societies. On both sides of the Atlantic vast and costly inputs of chemicals and technology into agriculture have produced vast and costly food surpluses. They have also produced land degradation and soil erosion, pollution, and a growing public anxiety about pesticide residues and additives in food. Rural communities are disintegrating and farmers, especially smaller ones, are giving up.

One of the virtues of Francis Blake's book is that he shows us an alternative, a form of agriculture that is non-violent towards the living soil, towards people and livestock; a husbandry which does not battle but collaborates with nature, which minimises the use of non-renewable resources and is, therefore, sustainable. The book is a message of hope for our society and for future generations, and especially for the farming community.

If organic methods result in some reduction of output, this will be a net gain – as taxpayers we now lose millions of pounds disposing of agricultural surpluses. And if, as experience shows, organic methods reduce output proportionately less than they reduce costs – if, for example, a ten per cent fall in output is accompanied by a reduction of twenty per cent in costs – then both taxpayer and farmer are better off. The case for non-chemical farming on economic, environmental, health and energy grounds is now overwhelming. The fact that virtually no government support is given to organic farming research and development is a national disgrace which owes much to the power of the agro-chemical industries.

The second Schumacher quotation highlights another quality of this excellent book: 'An ounce of practice is worth a ton of theory'. Francis Blake writes from personal experience. He started his own organic smallholding ten years ago and built it up to become a successful enterprise. A member of the Council of the Soil Association, he has been closely involved with the Association's Symbol scheme, its journal, and with its Organic Advisory Service.

This book is a practical introduction to the practice of organic farming, and a step by step guide to conversion to organic methods. There is a wealth of material on how to tackle each job, from soil care and manuring through to pest control and marketing. The book is well written, informative, and a pleasure to read because it is the product of a first-class mind, the stout heart of Francis Blake, and ten years of hard work on his farm.

George McRobie
President, The Soil Association

Introduction

CURRENT AGRICULTURAL PROBLEMS

Agriculture today is finding itself in increasing difficulties. It is being assailed on so many sides that it hardly knows which way to turn. The environmental lobby complains about pollution from pesticides, fertilisers and livestock effluent and about the rape of the countryside; the health-conscious are worried about the residues in their diet, and the tastelessness of food; the anti-marketeers point accusingly at the surpluses arising from the CAP (Common Agricultural Policy) and finally the great technological advances of recent years are seen to be causing rather than alleviating the terrible famines of the Third World. Farmers are desperate as their profit margins are squeezed and the policies which they are told to follow are continually being reversed.

Where is the way out of this predicament? Will any of the suggestions being considered by the authorities ever get to the root of the problem? Nitrogen tax, quotas, levies, taking land out of food production, more (but still secretive) pesticide legislation, provision in the grant structure for environmentally sound practices – however beneficial these proposals may be in themselves, they merely tackle the *symptoms* of our problems.

There is one solution that cuts a totally different path. It addresses all the problems currently facing agriculture, and so far it is performing well. It is organic agriculture.

When I started farming, I realised that it was a way of life. As I became interested in organic agriculture and then started an organic farm, I realised that it was more, it was also a philosophy of life. In order to appreciate the tenets of organic agriculture, you have to look behind its practical principles to this philosophy. Some people call it a 'holistic' philosophy, as it is concerned with the wholeness, the interconnectedness of life.

The principles of organic agriculture stem from this, and the practices follow on naturally. An organic approach to farming is therefore something that extends far beyond the farm gate, just as the effects of our actions do.

Imagine looking down on our planet Earth from a great height. You would see a small globe surrounded by a thin shroud of an atmosphere and covered in a paper-thin skin. Life, including ourselves, forms that skin, and its continued existence depends on an extremely narrow and precise set of parameters, found (as far as we know) nowhere else. Should they be upset, our whole world would collapse into unbelievable catastrophe. Luckily the time-scale of the Earth and its movements is far longer than we can experience, but the same principles apply in our own time-scale and at our own level. What happens to one part of this 'skin of life' affects all of it sooner or later, and to a greater or lesser degree. Destruction or damage is not confined in its effects to only one part of the earth; industrial pollution, deforestation, pesticide use and many more phenomena all bear witness to this fact, producing unsavoury ripples whose influence penetrates far beyond their own locale.

PRINCIPLES OF ORGANIC AGRICULTURE

Organic agriculture aims to be in harmony rather than in conflict with natural systems. This idea pervades all aspects of a farm, from how pests and diseases are controlled, through the treatment of livestock and the integration of the farm with the natural environment, to marketing, labour relations and health. The powers of nature are harnessed and developed to their fullest extent, rather than dominated.

Organic agriculture adopts an approach that minimises the use of non-renewable forms of energy. Chemical fertilisers and pesticides are either synthesised from oil or require large amounts of oil to extract and process them. It seems extraordinary that food production, the only method we have of actually producing energy, or rather of tapping the sun's abundant energy, has itself developed into a prodigious net user of energy. It is even more extraordinary that one suggested solution to the overproduction problem is to produce crops that can be turned into energy such as ethanol.

Organic food aims to be of optimum nutritional value. Varieties and methods of production are geared to this end. So too is processing, for having produced such quality food, no organic farmer wants to see it degraded, purified and processed, like white bread which needs additives to replace what is taken out in processing. The modern food industry is geared to its own interests – a food's storage life, impact on consumers and profitability – and not to those of the consumer which include nutritional value, health and safety.

The organic world strives to be localised. Local markets, decentralised systems of distribution and processing are sought. An old and much respected organic farmer tells the story of one time when he was visiting his local agricultural college. He got stuck in its rather narrow drive behind a lorry bringing potatoes to the college kitchens. It had met and could not pass another lorry, coming the other way, which was taking potatoes from the college farm to market! Our present system may be profitable for some, but it cannot be said to be efficient in a wider context.

Organic agriculture does not use artificial fertilisers or chemical pesticides; it is not, however, a low-input system. Contrary to popular opinion, these two statements are not at variance with each other. Organics may be a low 'external' input system, but, more importantly, it is also an optimum 'internal' input system. It may not be the most efficient in terms of output per acre or per man, but it is certainly the most efficient in terms of output per unit of input. In a finite world such as we have, this is important.

Organic agriculture does not pollute the environment. Residual poisons that stay in circulation to the detriment of other living things, including mankind, are not used. The waste products of conventional agriculture, that are so much of a problem in the current farming world, are not wastes to organic agriculture. Rather, they are the foundation upon which sound agricultural practices are based, for they return to the soil that which has been taken out.

Despite the fact that very little research has been carried out, and absolutely none by the Government, organic agriculture is emerging as a credible and entirely possible way of farming. What it could achieve with even a fraction of the research and development funding that conventional agriculture currently receives, is a subject fit for speculation. The mounting interest in organic farming, from all sides, is helping to fuel the demand for proper recognition, and is hastening the day when it will be regarded as a serious alternative that can make a very significant contribution to solving the current agricultural crisis.

This book is a small step along the way. It is designed for those who want to start farming and wish to do it organically, and for those who are already farming and are thinking about changing to organic. My experience is with farming organically on a small, but commercial, scale, and the book is therefore most relevant to that size of farming operation. For most of you who are contemplating the farming ladder for the first time, this bottom end of the range will probably be all that you can afford.

I do not cover the farming enterprises that only larger operations can manage, such as dairying and cereals. Instead I concentrate particularly on horticultural cropping, as this can offer higher returns on smaller acreages; and on the smaller types of livestock that are viable enough on the smaller scale and versatile enough to fit in with the horticultural side which is likely to be the mainstay of the farm.

With a subject as large and complex as organic agriculture, and as new as it is to most people, we can do no more than scratch the surface in a single book. I have therefore confined myself to describing only what is of particular relevance to our organic subject matter. Consequently I have assumed that you have a reasonable level of conventional agricultural knowledge, or that you have access to such information. Do not hesitate to use this book in conjunction with conventional books on agriculture and horticulture, which will supplement the information here.

1
The Organic System

It is a lot easier to say what organic agriculture is not, rather than what it is. This is partly because it is often perceived only as farming without fertilisers or sprays, and partly because any definition of it tends to be long and complex. Organic agriculture is a very different way of looking at farming, and this can make its concepts rather difficult to grasp, especially for those new to it.

SOIL

Soil is the one constant factor at the base of all farming operations. It is the beginning and end of all terrestrial life and the success and bounty of our agriculture therefore depend on it. From the soil springs plant growth, on which animals and man rely for food. The soil in turn relies on animals and plants for its food. It takes back what is left over, that is the manure and plant residues, to be recycled and used again.

Because of these facts, the organic farmer gives the soil his special attention. He particularly ensures that it is nurtured and fed with this 'soil food', for it is his most enduring and precious asset. In effect, the soil is the inheritance he holds in trust for future generations, and everyone knows that it is a farmer's duty to pass this on in equal or better condition than it was when he started. Of course all farmers should do this, but for the organic farmer it is part of his everyday practice. He feeds his soil for other reasons too. Most importantly, he expects the soil to feed his crops. This is the first great difference between organic and conventional standpoints. Soluble fertilisers feed the crop direct. They bypass the soil which, conventionally, is hardly more than a physical medium to hold the roots and keep the plant upright.

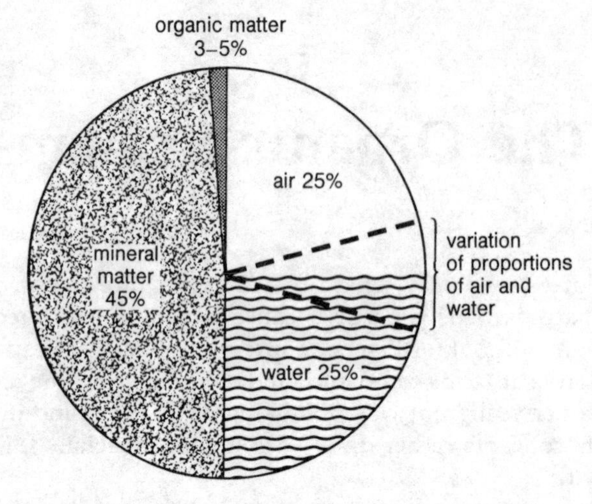

Composition by volume of a typical topsoil: note the surprisingly large proportion of air and water (although their exact proportions vary considerably), reflecting their importance as constituents of soil.

Biological Activity

Soil is certainly a physical medium, composed of rocks, stones, sand, silt and clay in varying proportions. We shall look at this in more detail later. What concerns the organic farmer is that soil is also a living entity – it positively teems with life. The number of micro-organisms (principally bacteria and fungi) in one small teaspoonful of soil is greater than the total number of people who have ever lived on this earth. Nor must we forget the other soil inhabitants, ranging from the almost invisible mites, through innumerable insects, nematodes, right up to earthworms and larger animals. Their health, vigour and balance make up what we call the soil's 'biological activity'. The aim of organic farming is to optimise this activity, because its product is the food which will nourish our crops.

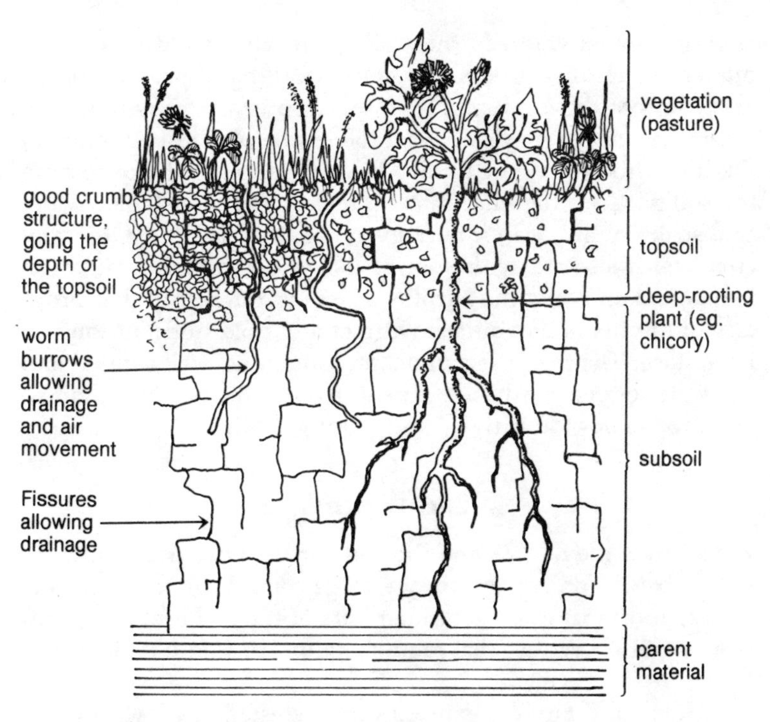

Profile of a living soil, showing good structure, root growth unrestricted by compaction or pans, and earthworm activity.

Organic Matter

The other, non-living half of the organic story in soil is the inert 'organic matter'. It is made up of simple nutrients (the plant food), carbohydrates (sugars, starch and cellulose, for example), proteins and lignins. The more complex constituents form what acts like, but actually is not, a single substance. It is black, amorphous, jelly-like or in minute particles, and has many different properties. It is called humus and it is very important.

All these components of organic matter are both the food on which the soil life lives and the products of their decomposition, and that of plants, when they die. In other words, the one depends on the other. They are in constant and dynamic balance and when we 'feed the soil' we are really adding organic matter to fuel this process, to get it working at peak performance, so that the food supply to the crop is at a peak too.

Organic matter plays a crucial role in the physical structure of the soil. This is why the humus is so important; it helps to bind the soil particles together into aggregates that are stable. Water can soak in and can drain away again, air can enter and circulate. It also creates a sponge-like effect so that some water is retained and only the surplus passes through. Another property of humus is its ability to attract and hold nutrient ions on its surface. The result is a moist environment, well aerated and not waterlogged, with nutrients at hand – the perfect environment for biological activity and plant growth.

Earthworms

What binds the humus and the soil particles together? How are they mixed? For the answer we must come back to one of the largest and most important members of the soil life, the earthworm. Gilbert White, that famous naturalist said it all in 1777:

> Gardeners and farmers express their detestation of worms; the former because they render their walks unsightly, and make them much work; and the latter because, as they think, worms eat their green corn. But these men would find that the earth without worms would soon become cold, hard-bound, and void of fermentation; and consequently sterile. Worms seem to be the great promoters of vegetation, which would proceed but lamely without them, by boring, perforating, and loosening the soil, and rendering it pervious to rains and the fibres of plants, by drawing straws and stalks of leaves and twigs into it; and most of all, by throwing up such infinite numbers of lumps of earth called worm-casts, which being their excrement, is a fine manure for grain and grass.

Gilbert White certainly summed things up well, and his last observation is what particularly concerns us here: the secret is in the worm casts. Charles Darwin, who studied earthworms in some detail, can enlighten us further. He wrote: 'All the vegetable mould over the whole country has passed many times through, and will again pass many times through, the intestinal canal of worms'.

The most important site where organic matter and soil particles are mixed is in the gut of the earthworm. Indeed, such is the activity and number of worms that Darwin calculated from one of his experiments that the production of worm casts is as much as 25t/ha per year. Perhaps that is not so surprising when one considers that the weight of earthworms in the soil under a pasture may be greater than the weight of livestock grazing on top. Certainly it amounts to a prodigious quantity of both worms and activity, all working towards the formation of fertile and well structured soil.

It has been said that if the right environment for earthworms can be created, it will provide the best conditions for plant growth. Perhaps that explains why organic farmers regard them so highly.

ROTATIONS

One of the corner-stones of organic farming is the use of rotations. A crop grown year after year on the same piece of ground is likely to run into difficulties very soon. As the seasons go by there will be a build-up of disease and the soil will have been increasingly exploited and depleted in the same way and at the same depth. This is obviously bad for both soil and crop, hence the need for rotations. Crop plants fall into several distinct groups – legumes, brassicas, roots, cereals etc. – and the organic farmer alternates these so that the gap between two similar crops is anything from three years upwards. Overall he wishes to achieve a balance through the rotation between deep and shallow rooting plants, heavy and light feeders, nitrogen fixers and consumers, and so on. A fertility building phase is also an essential constituent, to allow the soil and its life some undisturbed breathing space in which to build up their reserves again.

The same is true for livestock too; they may not deplete the soil, but each species grazes in a particular way, putting unbalanced pressure on the pasture. As with crops, disease and parasite build-up will occur, so the organic farmer should have more than one type of livestock to allow him the latitude in his management that will prevent these problems getting out of hand.

BIOLOGICAL CYCLE

Another corner-stone of organic farming is the careful use of all 'waste products'. This means returning back to the soil all manures and plant residues produced on the farm in the best form possible, that is, with minimum loss and maximum stability of nutrients. However, to say 'waste products' is to cultivate totally the wrong attitude. On conventional farms they may be waste, and a nuisance to boot, but on organic farms they are the means by which the soil is fed. They allow completion of the biological cycle.

Let us look at this a little closer. The food chain which culminates with humans, passing through soil microbes, plants and animals on the way, provides a direct link from the soil to ourselves. It does not stop there, however: it continues (or should do) from animals and man back to the soil microbes, to complete the cycle and to maintain and build fertility. This is a very important aspect of organic agriculture – it must be regarded as a *whole* system. All parts of the farm are linked in the biological cycle. What is done to one part affects all the rest in the end. No part can be treated in isolation. Thus, for instance, the organic farmer considers nutrient status and fertility build-up over the whole rotation; and cultivations are not just for making seed-beds, but also for encouraging an active soil life. Weed control is not confined to individual crops, but involves looking at all aspects of the farm, from rotation to cultivations, manure handling and what might come in from outside. The proportion of crops to livestock, and the effect of bought in feed or manure on other parts of the farm must all be considered.

PLANT AND ANIMAL HEALTH

Pests and diseases are all part of this interconnectedness. They are often the result of something wrong in the organic system, and are not merely a symptom of an immediately obvious problem. They can, therefore, tell us a lot too. Obviously a specific problem must be attended to without harming the rest of the system, but then comes the important part – an organic farmer will ask what the problem reveals about the rest of the system, why it has arisen and what can be done to prevent it from happening again. He recognises that such a problem is really a symptom of an underlying cause and it is that cause which he must find and tackle in the long run, using all his skill and knowledge.

It is generally recognised now that 'health' is a natural state. Disease is a consequence not so much of attack by a disease organism, as of an abnormal susceptibility to that potential attack. If you are down already, you are more likely to go under, whether you are susceptible because of stress, overdoing it, poor diet or some other reason. In our healthy, biologically active organic system, pests and diseases should not be a problem. There are obviously exceptions, but generally speaking organic farms are not bedevilled with pests and diseases. This is one of the hardest things for conventional farmers to grasp, for they are always worried about being without their armoury of pesticides. Basically they are not needed on an organic farm. Some naturally occurring pesticides can be used if necessary, along with biological control techniques, as they do not harm the rest of the system. These are usually sufficient.

BALANCE

Another important aspect of organic agriculture is that of balance. This is one reason why pests and diseases are not a significant problem in organic systems. In the natural environment all organisms are kept in check by the constraints around them. Predators and parasites, food supply and competition all maintain a dynamic balance.

Great fleas have little fleas upon their backs to bite 'em,
And little fleas have lesser fleas, and so *ad infinitum*.

This may seem a silly rhyme (originally coined by Jonathan Swift – quoted here from Augustus de Morgan), but there is much truth in it. Only when things get out of balance do populations explode, or die out. Similarly, pests and diseases have their own predators and parasites to control them. Unfortunately pesticides usually knock out the predators far more than they do their targets, besides disrupting the rest of the system that we have been trying to build up. The lush, unbalanced growth that soluble fertilisers produce in a crop makes it easy prey for pest and disease attack. Hence the need for that armoury of chemicals to counteract a self-generated problem.

YIELD

The yield difference between conventional and organic agriculture can range anywhere between 0 and 100 per cent. To be more specific is really quite difficult, because every crop and every system will produce different results. Little research has been done in this country, but some studies have been conducted in the USA and an overall average figure is usually somewhere between 10 and 30 per cent.

There are two factors that bear on these yield differences. First, do not forget the enormous amounts of research and development money that conventional agriculture receives, and, indeed, the enormous inputs from outside that are required to create these high yields. This fact perhaps highlights the efficacy of the organic system, which is an optimising system, and also what potential the organic system would have

(Opposite) The following beneficial insects prey on aphids. Ladybird: (a) eggs, (b) larva, (c) pupa, (d) adult. Lacewing: (e) eggs, (f) larva, (g) cocoon, (h) adult. Hoverfly: (i) eggs, (j) larva, (k) pupa, (l) adult. Others are ground dwelling and prey on all soft bodied creatures. Ground beetle: (m) and (n). Rove beetle: (o) and (p). Centipede: (q) and (r). The rule beside each illustration indicates the relative length of the adult's body.

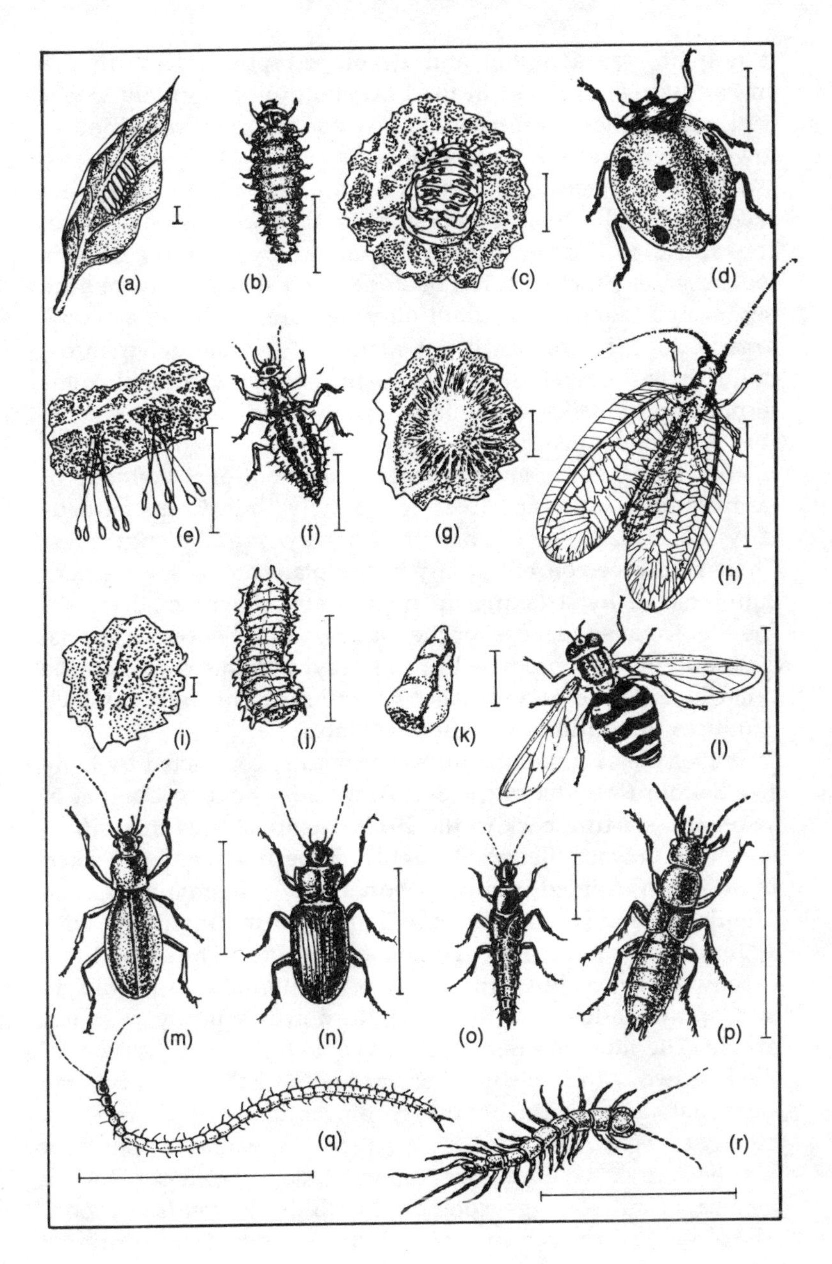
(a)
(b)
(c)
(d)
(e)
(f)
(g)
(h)
(i)
(j)
(k)
(l)
(m)
(n)
(o)
(p)
(q)
(r)

if only it was accepted and developed properly. With the embarrassing surpluses in the EEC standing at roughly 15 per cent, organic agriculture could make an important contribution towards rectifying the discrepancy.

The second factor is nutritional quality. Analyses of the nutritional value of the products of the two systems reveal considerable differences too, this time in favour of the organic food. Again, little work has been done on this subject, but the few comparisons of any credibility that are available, all show organic food to contain larger amounts of protein, vitamins, minerals and overall dry matter than their conventional counterpart, and smaller amounts of nitrates and free amino acids (the less good constituents).

Why is this? It is simply because conventional methods of farming push crops and livestock to their limit. Crops particularly are literally force fed with a narrow range of nutrients. Their excessive concentrations in the plant are to some extent counteracted by it taking up more water. Even so, there are inevitable consequences for the make-up and hence nutritional status of the food produced in this way. Organic crops, on the other hand, get a balanced and natural nutrient supply, which produces the optimum nutritional status.

One of the original organic experiments, conducted by Lady Eve Balfour (founder of the Soil Association) over a number of years and starting back in the 1940s, demonstrated this difference well. It was called the Haughley Experiment and consisted of one farm divided into three equal and self-contained plots. The first was mixed livestock and arable, run organically with all food, stock and manure generated within it. The second was exactly the same, but with all the conventional aids (fertilisers and sprays) added on top. The third was entirely chemical arable. The most interesting result came from a comparison of the first two. After they had become well established, it became clear that although the organic crop production was lower, the efficiency of its use by the livestock was some 10 per cent higher. It took 10 per cent less organic food to produce the same amount of milk or eggs, compared with the conventional food. The lower yield was compensated for by the higher nutritional value, and without the expense of all the conventional inputs.

PESTICIDES

Let us now look at the two main ingredients that organic farmers do not use, and the reasons why they are incompatible with organic principles and practices. Pesticides destroy the predators of pests along with the pests themselves. Indeed, these natural checks and balances are often more affected by pesticides than are the target organisms, because of their slower life cycles and dependence on the pest as their food source. This means that the pests are more likely to cause greater problems in the future than they did before.

Pesticides are often persistent; they do not go away. The very short life cycle of many pests means that they can quickly build up resistance to these chemicals. A vicious circle then develops wherever more pesticide needs to be used in order to have the same effect. This has an even more disruptive effect on the rest of the farm ecosystem, which becomes more out of balance. Other problems that were previously kept in check may then become prominent and need chemical treatment, and the vicious circle widens and deepens. Fungicides and herbicides too are lethal to other organisms, especially in the soil.

Not only do pesticides disrupt the ecosystem of the farm, they also pollute and endanger the countryside generally, and can work their way into every corner of the globe. It is a well-known fact that DDT has been found in the penguins of Antarctica, far from anywhere that the pesticide has been used. The organo-chlorine insecticides have been implicated in the declining populations of birds of prey. This is a classic example of species at the top of a food chain being affected: insects are eaten by small mammals and birds, who in turn are eaten by the birds of prey, and with each step up the chain the insecticide residues become more concentrated. The levels may not be lethal, but their effects are seen in other ways, like thinner egg shells that break before hatching.

We humans, at the top of our particular food chain, pour tons and tons of pesticides onto its lower levels. Ten years ago, mothers' milk in the USA was declared unfit for human consumption because of too high pesticide residues. The situation has eased somewhat since then, due to the ban on the use of DDT in agriculture, but the residues have by no means been entirely eliminated. Part of the reason for this might be that enforcement is only voluntary in this country, but their effectiveness encourages demand on the black market. Also DDT is still widely used in the Third World, much of whose produce comes straight back here.

DDT is not the only offending pesticide, it just happens to be a particularly fine example, being very persistent and having been very widely used. There are many others that are equally persistent and often more insidious in their effects. Aldicarb is one which has caused considerable trouble in the USA, not to mention India – the Union Carbide factory at Bhopal was producing it. In Britain, BHC or HCH (lindane) is no longer used for the annual sheep dipping against scab, partly because the French had higher standards that did not allow our short (two week) withdrawal period before slaughter. It is a sobering thought that BHC is detectable in the ewe's milk three months after dipping.

If one ignores all the horror stories about the dangers to both human health and the environment, there still remain some disturbing facts about these pesticides and the multinationals that produce them. The Pesticide Safety Precaution Scheme (PSPS) is a voluntary, self-regulatory system which is in the process of being updated. The data on which it is based is secret, and cannot therefore be validated by an enquiring public. Two years ago some of the laboratories in the USA, where many of the safety tests on pesticides are carried out, were found to be producing falsified results. These results came out in favour of their safety standards, when the reality was far from that. How can one trust these companies or their supposedly safe products when the wool is being secretively pulled over our eyes?

FERTILISERS

Soluble fertilisers may be food for plants, but they are not food for soil life. They bypass the soil life and go straight to the plant. If the soil receives no other nutrients, then its biological activity will gradually die away. The fertilisers also inhibit and can even actually kill parts of the soil's fauna and flora. Some potash fertilisers, for instance, are lethal to earthworms. In this way, the soil's fertility and biological activity can be very much reduced and it may take several years to regain its natural vigour and health.

The limited diet that soluble fertilisers present to the plant has a considerable effect on it. Too much nitrogen causes a lush, watery growth. Too much potash causes thin cell walls. Both these make the plant susceptible to pests and diseases. The concentration of these fertilisers in the soil solution means that other, lesser nutrients tend to get so diluted that they are not absorbed by the roots in sufficient quantities. Nutritional imbalances then occur which can have serious consequences not only for plant growth, but more importantly for the animals (or humans) that eat it. Mineral deficiencies can follow, for example, hypomagnesaemia in cattle grazing land that has received potash fertilisers.

Fertilisers move in three different directions; downwards, sideways and out of the farm gate. In fact, only about half of the fertiliser is used by the crop. The first movement, downwards, is perhaps the most dangerous, but we have not yet seen its full implications. Nitrate fertiliser seeps down into the deeper layers of the soil extremely slowly, until it eventually reaches the ground water. It takes twenty to thirty years to get there, which means that what is now coming out of the boreholes reflects the fertiliser use of the 1950s and 1960s. Compared to the present day, that was minimal, yet in some areas of the arable eastern counties of England, nitrate levels in borehole water are already beginning to exceed the EEC recommended limits. The tip of the iceberg is only just becoming visible. As more shows, the water authorities will be faced with an ever-increasing bill to make our water safe to drink.

The fertilisers moving sideways end up in our rivers and water courses, creating all sorts of ecological problems. First they stimulate vigorous growth of the water weeds and algae, which then use up so much of the available oxygen that other forms of life, like insects and fish, are suffocated. The water looks like pea soup and cannot support life. Phosphate and nitrate fertilisers are the culprits.

Fertilisers going out of the farm gate are in the food that we eat. The intensively raised plant tends to have more fertiliser nutrients knocking around inside it than its growth can usefully cope with. Nitrates, particularly, are present, and leafy vegetables contain the most – up to 70 per cent of the nitrates we consume may be from this source. The danger of too much nitrate in our diet is that it can turn into nitrite. This can be dangerous in itself, especially in babies (the blue baby syndrome), as nitrite blocks the oxygen exchange in the blood. However, nitrite can also react in the body to form nitrosoamines, which are a group of compounds that are highly carcinogenic. There is mounting evidence that nitrates can react with some pesticide residues to form similar compounds. Some medical drugs are also implicated, like aminophenazone for rheumatism.

PRINCIPAL GUIDELINES

Let me end this chapter by quoting from the first page of the Soil Association document *Unified Organic Standards*. It is the guideline for all organic producers in this country and sums up neatly the salient points discussed so far:

The development of biological cycles involving micro-organisms, soil fauna, plants and animals is the basis of organic agriculture. Sound rotations, the extensive and rational use of manure and vegetable wastes, the use of appropriate cultivation techniques, the avoidance of fertilisers in the form of soluble mineral salts, and the prohibition of agro-chemical pesticides form the basic characteristics of organic agriculture.

At the very least, you can now consider organic agriculture without any illusions. If you feel in sympathy with its aims and methods, you have taken the first step towards becoming an organic farmer.

2
Going Organic

Before discussing the practical aspects of going organic we should first take a look at the world in which an aspiring organic farmer will find himself.

OFFICIAL SUPPORT

Fifteen years ago, when I was at university, organic farming was dismissed as 'muck and mystery'. It was despatched in less than five minutes of the whole three-year course. Five years ago the situation was not very different. Since then things have changed considerably, but not as much as the organic movement would like.

We have been pressing for the Ministry of Agriculture to do more than pay lip service to organic agriculture, which is all that they are doing at the moment. Perhaps we should be pleased with even that progress, for at least lip service does indicate some recognition. Bureaucracy, the massive lobbying power of the chemical multinationals and a loss of MAFF's own direction have all contributed to this stalemate. However, progress is inevitable and the more pressure that can be brought to bear on the situation from those who want to eat organic, and particularly from those who want to farm organically, the sooner it will come.

At the moment there is no one 'official' body to turn to for advice. ADAS (Agricultural Development and Advisory Service) has no competent organic advisors, although some are getting interested as the pressure on them mounts. They have appointed someone in each area to oversee matters, but they certainly have no research or experience back up. There has been no research and development within the Ministry, ADAS or academic establishments in the past, although recently a few small projects have been initiated. The need is clear, as indeed is the precedent. In both Europe and America there is considerable interest and financial investment in organic research, and in some countries there are even government organic advisors too. Britain lags behind and her lack of foresight is beginning to show, as the agricultural scene becomes increasingly chaotic.

If the Government is not doing very much, then at least farmers and the general public are. Farmers, besieged on all sides (as we mentioned in the last chapter) are looking around in increasing numbers for alternatives to their present predicament. Different crops, co-operative marketing, intensification and even 'leisure' crops may help some. But others may be disillusioned as well as desperate, and for them the organic way may provide a possible alternative.

PUBLIC SUPPORT

The public are voting with their feet, and their purses. Media coverage is reaching almost astronomical proportions. The picture seems relatively rosy for the new entrant into organic farming, but a more accurate picture emerges in a comparative context. Less than one tenth of 1 per cent of agricultural land is currently farmed organically. In France, which is the country with about the highest proportion, it is more like 1 per cent. In virtually all European countries, organic development is far more advanced than it is here.

In this context, the public who are buying organic are still, relatively speaking, a very tiny minority. This seems to indicate that despite all the tremendous publicity and interest, it takes an enormous shift of public opinion to make a fairly small shift in 'activity', in this case buying organic. All the good signs are there, but it must be remembered that public inertia is a powerful force. Between this, the sceptics (of whom there are plenty) and the powerful chemical and food lobbies, the spread of organics faces no small opposition. Perhaps the fact that the organic movement is so tiny means that it can only grow.

ORGANIC ORGANISATIONS

So much for outside the organic movement; what is the organisation within it? Forty years ago a group of far-sighted people came together because they were worried about the way agriculture was heading, and the consequent implications for human health and the environment. They formed the Soil Association. Today it continues its original aims of promoting organic agriculture, educating the public and providing a philosophical umbrella over the movement.

The Soil Association

Everyone interested in organics (farmers and consumers alike) should be a member of the Soil Association. Its role has changed somewhat in recent years, from the specialist emphasis on research and promotion towards more general alignment with and education of consumers. In that capacity it runs educational campaigns and also administers a licensing system for organic producers, to safeguard organic integrity for the public. This is the Symbol Scheme and it is based on the organic standards already discussed. The Symbol is awarded to those farmers who are considered to be farming up to those standards. All aspiring organic farmers should be aiming at getting the Symbol and all existing organic farmers should have it, if they are worth their salt.

International Federation of Organic Agricultural Movements

Throughout the world, organic standards are set and co-ordinated by IFOAM, the International Federation of Organic Agriculture Movements. They are applied in Great Britain by a committee consisting of representatives of all the organisations in the organic movement. The standards encapsulate our understanding of organics and its application in the field, and so underpin much of what is said in this book.

Producers' Organisations

There are two producers' organisations: the Organic Growers Association and British Organic Farmers. Both are quite young, which indicates just how far behind our European colleagues we are. OGA was formed in 1980 and BOF in 1983. Nevertheless, both organisations are growing rapidly and their stature in the organic movement belies their age. They work very closely together and produce a joint journal, *New Farmer and Grower*, which is full of practical and theoretical articles, advice, tips and much more besides. They organise farm walks, seminars and conferences, and operate an information service for members. Obviously membership of one or both organisations is a must for anyone contemplating organic farming, as the information you will gain, the contacts you will make and the farms that you should visit will all be vital in the process of change that you and your farm must undergo for a successful conversion.

Elm Farm Research Centre

The next important organisation, you will be glad to hear, does not need joining. It is Elm Farm Research Centre, and is the only organic research body in the UK, privately funded and based on its own organic farm near Newbury. However, you will almost certainly be sending them some money, not for membership but for their organic soil analysis service. The ADAS soil analysis is next to useless from an organic point of view as it only indicates what fertilisers are needed. The organic soil analysis offered by Elm Farm has some fourteen different readings, which gives much more information not only about nutrient status and reserves, but also about the biological activity of the soil, structural problems and heavy metal levels. This analysis will probably be required for your Symbol application, at least if there are any doubts about your soil.

Elm Farm Research Centre is closely involved with the Symbol Scheme, investigating any problems that arise in applications, analysing produce or outside sources of manure, testing new products and so on. On the farm field trials are carried out, different varieties, machines and methods are tested, and a host of other experiments are done. The centre then puts its experience, together with research from elsewhere, into practical handbooks which are available to buy.

Organic Advisory Service

One aspect that has been sadly and noticeably lacking until very recently is an advisory system for organic producers. In 1986 this was remedied and the Organic Advisory Service was born. Run mainly by experienced organic farmers who are willing to spare some time, and a few full- and part-time consultants, the Service is co-ordinated and overseen by the four organisations mentioned above. To ensure the quality of the advice, potential advisors are screened and training and updating courses are regularly arranged.

Making use of this service, which unfortunately is not free, really is necessary. This book should ensure that your initial mistakes will be fewer, or at any rate less serious, but a newly organic or converting farm, combined with the inexperience of a newcomer to the system, are guaranteed to throw up problems and cause blunders. It is therefore essential to have someone experienced at hand who can both guide the general learning process and help with trouble-shooting in an emergency. It is also important to ensure that the conversion of the farm is carried out in the most efficient way, and the organic advisor is the best person to oversee this. I have been to see many farmers or growers with a view to assessing them for the Symbol, who had not yet contacted anyone because they knew that they were not fully organic. As a result, they had often lost a year, or even two, of organic status due to mistakes which could have been avoided by following the right advice at the right time.

Other Organisations

There are many other organisations in the organic movement. Prominent amongst them is the Henry Doubleday Research Association for gardeners, the McCarrison Society for medics, and the International Institute of Biological Husbandry for academics. Others are given at the back of the book, but one more should be mentioned here: the Biodynamic Agriculture Association. Based on the teachings of Rudolph Steiner, biodynamic agriculture is a sort of organic 'plus'. It is too involved to discuss here, but it is widely practised in Europe and is highly regarded. First, however, you have to be organic.

EXPERIENCE

If you are contemplating farming for the first time, you must pause to consider what is involved in the business. It is said that farming is a way of life. It penetrates every part of your activity. You live on the job, you might be up at all hours of the night, many working days will extend long into the evening and weekends will almost be a thing of the past. Not only that, but the skills and knowledge required to run the farm effectively make a long and awe-inspiring list, from businessman and salesman to vet, weather forecaster, scientist, agronomist and many more. A farmer is a Jack of all trades, and a master of most.

The prospective organic farmer must be aware of this to an even greater degree because his work is likely to involve even longer hours, his knowledge must be deeper and his skills will play a more critical part. You cannot cut corners by using chemicals, nor do you have recourse to the welter of advice from all sides that other farmers enjoy. Only a foolish person would set about procuring their own farm without first acquiring as much experience as possible on other organic farms, learning by other people's mistakes and gaining skills and knowledge at other people's expense.

Acquiring Skills

Another organisation that might help you to gain experience is Working Weekends On Organic Farms (WWOOF), which arranges for city dwellers and other interested people to spend certain weekends working on member organic farms. It is a very good way of getting experience from scratch, but most of the participating farms could be classed as smallholdings and are therefore less useful beyond the basics.

Another and more effective way of gaining experience is to live and work on an organic farm of roughly the same scale and type as you have planned, for maybe a season or even a year. Many organic farms operate just such a system (with students working for board and pocket money) and are equipped with caravans or living-in facilities. Foreign students take particular advantage of this. Indeed, it is a very good way to learn, especially if you approach it with the right enquiring mind. You will probably learn about the long hours and hard work as well!

There are really no courses that you can attend in Great Britain. Agricultural colleges and universities lag behind our continental colleagues and provide nothing at present, although within the next few years the situation will probably change considerably. Worcester College of Agriculture, Aberystwyth and Bath Universities all have plans in the pipeline at varying stages of development.

The Agricultural Training Board can be a very useful source of information and experience. Local groups decide the subjects on which to have courses, the instructors being skilled local farmers, and several have chosen aspects of organic farming. Otherwise, their ordinary courses provide excellent practical training in many diverse subjects ranging from hedging to mechanics.

In addition, it is just as important to gain skills and knowledge in the ordinary, non-organic aspects of farming. In fact, they are the base upon which the extra skills required by organics must be built. Stockmanship, husbandry, costings and economics, mechanics and so on, not to mention the more nebulous qualities like attention to detail, resourcefulness, self-discipline and tenacity are all absolute necessities if you intend to farm well, and particularly so if you intend to farm organically.

Lastly, learn from books. There are few practical books about modern organic farming, but there are many reference books and ones about conventional agriculture that will be relevant and are worth having around. There are also many books written in the first years of the organic movement that hold a wealth of wisdom, even if they are not of actual practical use today. These give an excellent grounding in the principles and philosophy of organics, which are both important areas for the novice.

STARTING OUT

Let us now turn to the practicalities of starting, assuming that you have under your belt all the experience you think you need. There are basically three possibilities: buying and starting from scratch, renting, or converting an existing farm if you have one already. It may be that once bought or rented, the farm will have to be converted anyway. The business of renting is very much the same as buying, the same general criteria apply, and it is just that the recipient of your money and the amounts needed may be different. I therefore do not propose to say any more about renting, but will leave you to make your own comparisons. It may, nevertheless, be a way for someone with less capital to get into farming, if the rent is not prohibitive.

BUYING

A welter of factors and considerations must first be assessed, beginning with the amount of money you have available. How much have you got? Out of that, how much should be spent on buying the holding, how much set aside for stock and machinery and how much left for the initial working capital and income, and for contingency plans? From this information, how much do you want or need to borrow? Only by working out plans and budgets for the first few years will this become clear. But do remember to budget very much on the low side and to give yourself something to live on in the first year.

Many people believe that the land should not be expected to carry the burden of paying for its purchase price. This, they say, puts pressure on the farmer to exploit his land over its limits and hence encourages the intensive methods of today. Looking at the management and investment income statistics of agriculture, one can certainly see why: the figure varies around 3 or 4 per cent, which is not much return on all the capital and hard work put into a farm. If the farm has to borrow too much on the open market to finance itself, problems and even disaster might well ensue; many small farms, now amalgamated and lost, can testify to this.

Be warned, and do not over-stretch yourself at this early stage unless your plans are watertight and crystal clear. It is a well-known fact that it is very difficult to get into farming if you do not have money. Unfortunately organic farming is no exception and the situation may, if anything, be worse; certainly this is often asserted, perhaps with reason as there is no recourse to exploitation. The only mitigating factor is that many organic farmers are prepared to accept a lower income, at least initially.

Choosing a Location

Having decided what money is available, you can then look at what it might buy. Size, quality, location and type of farm all enter the equation. A hundred hectares of moorland in Scotland might cost the same as fifteen hectares of grade one land in East Anglia. Where do you want to live, and what sort of farm do you wish to have? Somehow your finances and your preferences have to come to an arrangement together. If you want an extensive livestock system, choose somewhere higher and wetter; it will also be cheaper. Intensive market gardening, on the other hand requires less land of better quality, and is therefore more expensive per hectare.

To go into detail here about the various agricultural areas in Britain is to go beyond the scope of this book. Variation within them is also so diverse that generalisations become almost meaningless. Local ADAS offices are the best places at which to direct your enquiries when you have narrowed down your choice. They will soon tell you what you need to know.

The local Meteorological Office is the best source of guidance on matters of climate. The main factors are obviously sunshine, rain and temperature, but do not forget their distribution through the year, and also wind and general exposure, including altitude. Although the area may have a general pattern, certain parts are likely to alter considerably from this due to odd topographic features, such as a range of hills causing a rain shadow or certain valleys making frost pockets. If the Meteorological Office cannot go down to such detail, then

ADAS should be able to, so add it to the list of things to ask them.

Further information can be gleaned by looking around you at what others are doing in the area. However, do not necessarily be guided too much by local practices. For example, the vale of Evesham is a classic horticultural area but, because of that concentration of crops, it would not be the best place to start an organic horticultural enterprise, even if you could afford it.

On the small scale it is often possible to adopt methods and crops that are otherwise unsuitable for the larger farms in a certain area. This allows greater versatility, especially in the poorer agricultural areas of the country. For instance, many fine examples exist in the hills of Wales, an area that attracts aspiring 'drop-outs' because of its beauty, isolation and particularly because it is often all that they can afford. The indigenous farms are mainly sheep, beef and some dairy with virtually no arable cropping, because the land is high up, wet and of poor quality. The organic farms, however, manage to produce a wide range of winter and root crops; indeed, West Wales constitutes one of the principle areas for organic carrot production in Britain. This seems totally at variance with the conventional scene, but is made possible by small scale adaptability, the greater resilience of organic soils and the absence of carrot root fly at these altitudes – an important consideration for organic carrots.

ASSESSING SOIL

You will need to know a bit more about soil in order to make an informed choice about any particular farm.

One of the extraordinary things about British soils is their enormous variation, partly the result of the grinding, mixing and moving that went on in the ice ages of the last half million years, partly because of other forms of erosion and weathering of the ground rock. The physical part of the soil is easily divided up into categories by size of particle, but the distinction between them involves a lot more than just size. Each size range has distinct properties which affect the nature of the soil from an agricultural viewpoint.

Stones and Sand

Stones, and anything larger, obviously make cultivation more difficult. They may have some influence on drainage and also on structure. Sand creates an open and free draining texture as the particles are large and irregular enough to allow the passage of water. Sandy soils have very few cohesive properties and are 'light' to work. They therefore dry out quickly and need plenty of organic matter to maintain a more stable structure.

Silt

Silt is the next particle size down. The particles are too small to have the physical properties of sand – they just run together and block up all the holes and pores – and are too large to have the chemical properties of clay. Silty soils are therefore difficult to work, their structure is virtually non-existent (without assistance) but they are often very fertile.

Clay

Lastly clay, composed of the smallest and probably the most important particles in the mineral part of soil, is also the most complex in terms of both its action and its effect. The clay particles are plate shaped and are so small that they are chemically active. They have 'colloidal' properties; that is, they have the power of sticking to other substances, including chemical elements or compounds such as nutrient ions. The significance of this is profound on two counts:

1. Clay particles, although so small, can form larger groupings that are more or less stable, depending on the type of clay and other considerations. Water can penetrate into these groupings and be held against drainage, as well as passing by outside them. Furthermore, as the soil is wetted and dried, the clay particles expand and shrink, causing cracking. This is also done by the winter frosts as the soil thaws and freezes. So, soil structure can be good and drainage facilitated if the right cultivations are done at the right time. The margin, however, is fairly narrow, clay soils being sticky and plastic when wet, and hard

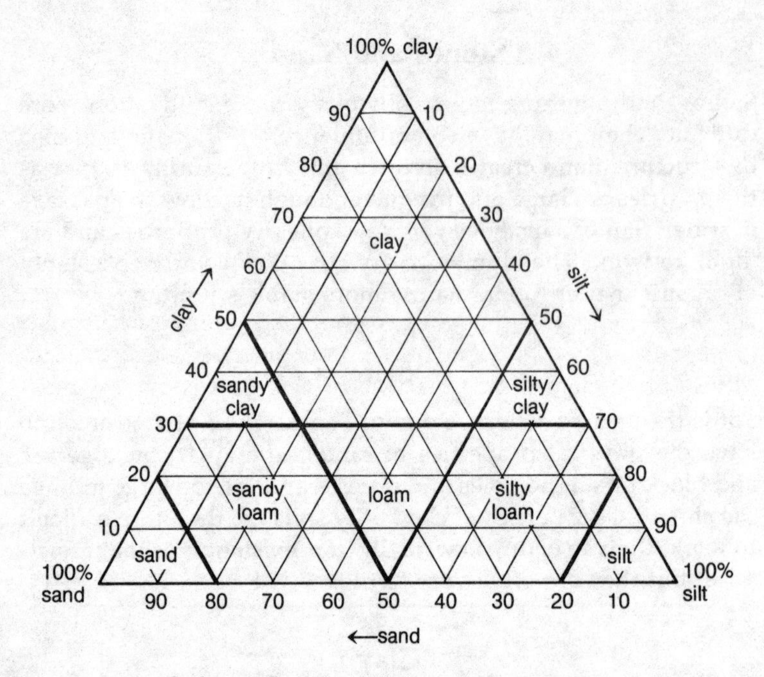

Soil triangle showing calculation of soil type.

and blocky when dry. In the subsoil, drainage must be safeguarded too, which means subsoiling when conditions are dry enough to ensure that cracking occurs.

2. Clays do have their advantages too. Their ability to hold nutrients on their surface, which gives them the same exchange capacity as humus, points to potential for good fertility without the need for such large amounts of organic matter as the more sandy soils require. Needless to say, clay soils benefit from organic matter in other ways; to help make them more workable, or to lighten them, for example.

Loams

It is seldom that you will find a soil composed of only one constituent. Most British soils are mixtures of all three, sand, silt and clay, and although one may predominate, its behaviour is tempered by the presence of the other two. These soils are called loams and they make up three quarters of the soils in Britain. The mixture of large pores from the sand and the smaller particles that then fill them up, ensures that the biological means of creating good structure are of prime importance in the vast majority of British soils. These consist of encouraging earthworms and good root development, to produce water-stable aggregates.

Peat

A further group of soils that occur in Britain are organic or peat soils. Do not confuse the term organic here with what we are talking about in this book; it simply means that the base material is of organic origin rather than mineral, because for various reasons the conditions during its formation were anaerobic and the vegetation was unable to break down. These soils tend to be too acid for earthworms to survive in them, so they need considerable work on them before they can become productive. After that, they can be extremely fertile, as in the fen peats, for example.

Soil Suitability

Without wanting to over-simplify, whatever soil types confront you in your short list of farms, they will all benefit from organic methods. What needs closer attention is the soil's suitability for the cropping regime you have in mind. For example, the heavier and more clayey soils are likely to be colder, wetter and will crop later. They will therefore work better under less intensive rotations, with long ley breaks, growing cereals, where earliness is not important and cultivations can be well timed. The sandier, lighter soils, in contrast, make good horticultural land, allowing frequent working unhampered by weather, with the proviso that the organic matter levels are maintained at adequate levels. There are very few pure silty soils, with no structure, that really need to be in permanent pasture. More often they have other constituents as well and can make good land if due attention is given to the structure; they need the organic matter of sandy soils and the timeliness of cultivations of clay soils.

WALKING THE FARM

Once a short list has been drawn up according to your preferences of location, type, size and price range, it is time to look at these farms on the ground. What indicators can tell us about the state of a prospective farm? You will want to find out its previous use and present performance. You must therefore glean as much as you can from the vendor about yields, cropping history, fertiliser and spray regime, manuring policy, reseeding programme, stocking rates and so on. Much of this will later be required for your Symbol application. If you are hoping to convert the whole farm at once, the less artificial fertilisers and sprays have been used in the last two years, the better. High applications of either will require that the land be converted under a ley for at least two years. As your main enterprise is likely to be arable or horticultural, this could be a determining factor against the farm, unless you have other plans to get you over those first few seasons.

You will need to assess whether the farm is going to be suitable for the crops you have in mind. Frost, exposure, workability, pH, slopes and drainage will all have an influence. In addition, you will want to know about the quirks of individual fields, their earliness, wetness, stickiness, performance and pest and weed problems.

Whenever you go to inspect a farm, keep your eyes open. This is a necessity for all farmers and especially organic ones, so develop it from the start. Look about you and assess the holding in general. What state is it in? Do the crops and livestock seem healthy? Does the place look a mess, uncared for and worn down? You will need to decipher how much is due to a worked out, over-exploited farm, and how much to bad management. Although this may not be easy, the position should become clearer as your inspection progresses. If over-exploited, it might require a long and therefore costly recuperative programme to coax it back into shape, but probably no more than if it was suffering from too many chemicals. Organic methods are the ideal means.

Always take a spade and use it in each field to evaluate the depth of topsoil, the depth of rooting of crops and pasture, the number of earthworms and any other features that are apparent. Mats of organic matter at the surface, mottled layers of soil and hard pans all suggest poor structure and drainage and lifeless soil. Conversely, deep, vigorous, well-branched root systems, lots of worms and crumbly tilth suggest a healthy soil that might be well on the way to organic acceptability.

Worm Test

A very simple test (called the worm test) can easily be done on the spot to find out roughly the physical constituents of a soil. It has nothing to do with earthworms, but looks at the plasticity of the soil when wet. This is what you do: scoop up a small but representative sample of soil, enough to fit into an egg-cup, and work out all the coarse bits, stones, roots, etc. Then, treating it like a small load of cement in your hand, add water bit by bit, mixing all the time until the consistency is smooth and plastic. The most convenient source of moisture in the middle of a field is your own spit, so make sure that you are salivating well! Now start forming it into various shapes; first a cone, then roll it into a ball, then a worm, just like children do with Plasticine. Then try bending it right round in a circle. The further you get, the more clay there is in the soil.

cone: sand
ball: loamy sand
sagging ball: silty loam
worm, cracking when bent: loam
circle: clay

Beaker Test

The worm test will give you an idea of the physical nature of the soil and whether it is suited to the sort of enterprises you have in mind. Equally important, however, is its quality, its progression towards full organic health, its 'heart'. If a soil is good for worms and they are flourishing and producing good water-stable aggregation, we can expect that it will also be good for farming. One quick way of measuring this is the 'beaker test'.

Take a representative sample of soil, preferably on the drier side, and sieve off the loose fine particles with a kitchen sieve. Half fill a glass beaker and pour in a dessert-spoonful of what remains in the sieve. Let it soak for a few minutes and then swirl it round. Watch what it does. If the soil stays fairly intact and settles without going cloudy, then it is water-stable and can be expected to be in 'good heart'. If, instead, it goes cloudy, then you and the worms have got work to do; alternatively, it is too sandy or has too little clay (less than 10 per cent) to hold a structure for the purposes of this test. You should be able to find out easily enough if the soil tends towards that extreme.

Weeds

Something else that can tell you a lot about the state of the farm, and not just about its state of neglect, are its weeds. Weeds are nothing if not adventitious. They spring up wherever there is a niche for them, whether formed by cultivation practice or by deficient soil or other factors. Weeds can be regarded as indicators for these various situations. Familiarising yourself with this aspect of plant lore will help to further your understanding of the workings of biological processes. A few examples might help at this point, but they can only be very general.

Acid and alkali soils both have their typical plants, but of more interest in this context are factors like drainage. Horsetail (*Equisetum*) is a classic plant of acid clay soils in need of drainage; redshank also indicates lack of drainage. Corn marigold and corn spurrey both prefer light acid soils; sorrel, dandelion, docks and many others prefer heavier, acid soils. The penetrating roots of coltsfoot and couchgrass prefer heavy clay soils which they help to aerate. This introduces the very interesting concept of the 'corrective' nature of so many weeds, for instance their ability to accumulate from the soil those minerals that are lacking.

Weeds can reflect the intensity of use of land. When I first started my market garden, the most common weeds were red-shank and camomile. Gradually, as the pressure on them mounted, these were replaced by fat hen and annual meadow grass, and these in turn by groundsel and annual nettle. Chick-weed, I might add, has stayed persistently throughout; it is a weed indicating high fertility, for all its other faults. Growth habit will also obviously reflect the fertility present in a soil. For instance, if groundsel grows rich and tall, all well and good, but if it is small and too eager to flower, then fertility is lacking. The presence of leguminous weeds (vetches and medicks) would also suggest low fertility, particularly nitrogen shortage.

Fixtures

Look around you some more, this time at the fixtures of the farm. Perhaps the most important are the farm buildings. In what state of repair and how comprehensive are they? To what extent will they need renovating, modifying or adding to, for your purposes, and how much will that cost? Look also at the hedges, fences and gates and apply the same criteria; then at the water supply, and at the access to both the farm and individual fields.

Once you have satisfied yourself that the farm itself comes up to your requirements, then you can look to your own comfort – the farmhouse, if there is one. It all depends what level of comfort you think you can survive on. Some people can thrive on just the barest essentials whilst others require their luxuries about them. It also depends on what capital you can set aside for this non-productive aspect of the farm. Weighed up against that, however, is the amount of time that renovating or doing up the house might take, valuable time taken at the expense of the mountain of work piling up outside. Just remember that farm work is hard work, and after a strenuous day it is important to have at least reasonable comfort when you stop.

Other Considerations

The next thing to do is to take a break and pay a visit to the nearest pub. There you will hopefully meet your prospective neighbours and other locals, from whom you should be able to glean further information about the farm, about local conditions and also about your neighbours themselves! It is important to be on good terms with them all, for you never know when you may need to call on them for advice or help in the future, or worse, to negotiate after a spraydrift incident.

Now you can go home and consider all this jumble of information and make your initial decision. To help you, try and stand back and look at the farm as a whole. How closely will it be able to approach the organic ideal of a balanced, whole system within itself? Is it suitable for an organic farm? Looking ahead, how will it fit into any existing organic structure in the locality, such as marketing, co-operatives and support in your early days?

If things seem positive and you decide to take the plunge, you will have to sit down and work out in much more detail your capital budget. First include all the fixtures that need attention, repair or renewing (from water troughs and drainage to the house itself); some costs will be spread over several years, others will be at once. Then work out your 'running' capital, the cost of stocking and mechanising the farm. With capital or borrowings put aside for them, you can finally arrive at the purchase price you are aiming for.

Good luck! It will be a nail-biting time, but there will be plenty of homework and planning to keep you occupied, contacts to make, arrangements to finalise and lots to buy. There are two things I would suggest you do at this time, which are relevant to the farm's organic progress: the first is to seek out good organic advice via the OAS or other means, to help you start on the right path from the outset. Secondly, to complement this, have the soil analysed with the organic analysis service from Elm Farm Research Centre. If there are many different fields and many different soil types, several different samples should be sent. The information that this will give you, particularly if you repeat the sampling every year as you should, will be both fascinating and invaluable.

3
The Conversion

You have just walked out of an auction as the proud owner of your new farm and wish to convert it right away. You are lucky enough to be starting with a clean slate, although you will have the financial pressure both of starting up and of converting all at the same time. For that reason, it may not be possible to convert all of it at once, and some of the fields may need to wait and earn money with chemical arable rather than in the less profitable conversion ley, whilst you find your feet and get used to this new situation. If the farm's recent history has been chemical-free, your job will be that much easier and you will have much more latitude in implementing your chosen ideas. In any event, the processes of analysing the suitability of the farm for an organic system (which you will have done before deciding to buy it) and then deciding how to proceed towards it (done at the homework stage) are exactly the same as if you had been there for a number of years. You can compensate for your lack of experience of the farm partly by the homework that you do, and partly by being able to make a start unhampered by previous ruts.

Alternatively, if you have a small conventional farm, you will have been there long enough to know it well and your enterprises will be established and stable. It is likely that economic reasons have forced you to look at this possible alternative, or perhaps a health problem in your family has caused you to question your present practices. Your interest is aroused, you have made some initial enquiries that seem positive and you now want to look more seriously at what it might mean for you and your farm.

You will be aware, by now, that there is a great deal more to organics than just cutting out the chemicals and waiting for the organic premiums to start rolling. The effect on your farming system, especially one that is more arable than livestock, is likely to be dramatic.

THE FARMER

From this point, either going into organic farming or converting your existing farm will have the same major implications for yourself and your family. The first step in the conversion of a farm is the conversion of the farmer.

There are two requisites. The first is a change in your attitude and your thinking about farming. You will need to get away from the conventional approach to any problem, be it pest, disease, weed, fertility or whatever – that of seeking to dominate it by using outside inputs – and instead cultivate the ability to look at it in the context of the rest of the farm, to see behind the problem to what is causing its undue development, and to make appropriate changes and adjustments there. You will need to stop regarding each enterprise in isolation and look at the farm as a whole, for each part has an influence on everything else.

The second requisite is a belief in your own mind that the organic approach is the right one. In order to feel confident and positive about this new direction, so that you can carry it through even in the difficult times and against the barracking of your neighbours, then you have to have put your faith in it. An experimental attitude, merely leaving a field unfertilised to see what happens, is doomed to failure because it is neither organic, nor is the commitment there to do it seriously. Without that whole-hearted belief, there is no room for the development of either the pioneering spirit or the understanding that organic farming requires for its successful operation.

It is bad enough being isolated and under fire from neighbours, as might well happen, but it is far worse if you are at variance with your family too. Their support for this new organic direction is very important as the farm is so often a family unit in which all participate, and it at least gives you someone to turn to near at hand. Luckily this is usually the case.

THE FARM

The farm too must be suitable for conversion. Again, this assessment is in two parts: first are the physical aspects, its soil and climate. Certain conditions make a conversion difficult or impossible; for instance on reclaimed soils where the topsoil has been heaped up in enormous mounds for a long time, and its biological activity has been killed by the anaerobic atmosphere inside it. It takes many years to revitalise such a soil that has been so totally deadened. Another case which would be difficult or impossible is if the land has had undue amounts of sewage sludge or other additions, that have given it too high levels of heavy metals. These can cause problems in organic systems and for health reasons are not permitted in quantity in the soil. Unfortunately they take many, many years to disperse. The same might be true of some agrochemicals that have been used in previous years. Many are supposed to bind on to the soil and be rendered harmless, although they may stay there indefinitely. Others do not lose their potency or perhaps break down to equally poisonous constituents. Methyl bromide, the soil sterilant, is one such chemical which remains untouched through succeeding years.

Soil type is not usually a limiting factor, as long as the planned system is compatible with it. Deficiencies or structural problems may require long-term work to correct them, in which case the conversion may have to be delayed or extended. Similarly climate, although not a limiting factor, must be taken into account. An organic system, striving to be in co-operation with nature, must be suited to the conditions in which it has to operate.

It is obvious that many of the above considerations require more information about your soil than might be readily available. As with the process of buying a farm for organic production, so with converting an existing one, you will need to obtain this extra information. The only way is through the organic soil analysis done by Elm Farm Research Centre.

The second part of your assessment involves the farm activities and enterprises. Certain practices can not be converted, like intensive pig and poultry units, or monocropping, for they contravene the principles of organic husbandry. Certain others are difficult or impossible, like some horticultural crops that require special chemical treatment to make them viable. Still others are possible but pointless in an organic system, such as crops that do not have an organic market, or are processed out of existence.

Generally speaking, the more specialised the farm is, the more difficult it is to convert to a sound organic system. More and greater changes will have to be made to achieve the necessary diversity, which is likely to lengthen the conversion process. The adaptability of the farmer in such a situation is stretched to the utmost. Similarly, an all-arable farm will be more difficult to convert than a mixed or all-livestock farm. All-arable organic farms exist in Europe but none are yet established in Britain. On the smaller scale, purely horticultural units are common in Britain, but they inevitably depend to some extent on outside sources of fertility. If these are available it is an option that can be considered.

The next factor to assess is the financial state of the farm. The returns during the conversion process are going to be lower than normal, because the fields under conversion are going to be yielding very much below par until they regain their organic health and are cropping in the new rotation. Therefore any additional burdens like the servicing of large loans will make the situation particularly difficult. The time over which the conversion runs may have to be extended so as to spread its cost over a longer period.

There are actual costs too, as well as reduced income. Establishing the conversion leys, including the seeds mixture and any soil amendments that might be needed is one area. Another, and larger one, is the cost of any additions to livestock and machinery that the change requires. For instance, new weed control and cultivating equipment will almost certainly have to be bought, manure handling and harvesting equipment might need altering, and type and variety of livestock may well have to change too.

PLANNING

You can now look forward more definitely towards the ideal organic system that you want to introduce on to the farm. With your present situation, the organic principles and an inkling of your ideal aims, you can work them all in together to form a smooth transition towards full organic status.

A close inspection of all aspects of the farming operation must now be undertaken to find out where things are in conflict with the organic 'whole system' approach, and therefore what needs changing, adjusting, or even cutting out completely. Then a plan must be drawn up on the basis of all these factors that will facilitate this process of conversion in the context of the financial and practical situation of the farm.

The correct decisions made now will make the whole conversion that much easier. If you have not already done so, it is time to enlist the services of someone qualified and experienced enough to ensure that you make the right decisions; you may know (or think you know) what is best for your farm, but you are unlikely to have yet gained the crucial organic expertise and experience that is needed here. The full conversion plan is quite a technical and, in some places, theoretical document which you will be unable to compile properly on your own. Professional advice is therefore necessary, and you should not begrudge the cost.

It is extraordinary that such a dramatic change as a conversion to organic should ever be contemplated without first consulting the experts in the field. No other change on the farm, certainly of this scale, would ever be undertaken without having the feed or chemical company representative, or ADAS, looking over the farmer's shoulder and advising his every move. With organics, however, many people seem blind to the very complex nature of organic systems. Perhaps they are still hooked on the 'mystery' myth, and thoughtlessly launch some field into a supposedly organic experiment without so much as an initial enquiry into the current state of the art. The consequences are inevitable – failure.

In conjunction with your new organic advisor, the first part of the plan can be drawn up. This involves preparing a fairly detailed analysis of the current situation on the farm. You will already have looked at much of this, as it includes the standard farm data like acreage and layout, soil types and nutrient status, rainfall and growing season. It also includes the various enterprises and the inputs that they rely on – food, fertilisers, pesticides, mechanical operations, etc. This is the starting point from which the second part, the action plan, is developed.

ACTION PLAN

The action plan constitutes the conversion itself and therefore has to take into account a number of factors. These will be discussed individually so that you get a clear idea of what is involved and what you have to achieve. They sum up much of what has been discussed in previous chapters and put it all in a context that can be applied on the ground.

Soil Improvement Measures

The soil may need improving in two ways: structural and nutritional. Drainage and aeration are important factors in encouraging good biological activity. These should therefore be attended to, not necessarily by capital drainage schemes (although this is a sensible time to consider them if they fit in with other plans and budgets), but more by deep cultivation techniques like subsoiling, which will break plough pans and other compaction problems. Aeration and loosening of the soil can be helped and stabilised by establishing a fast and deep-rooting green manure after appropriate cultivations.

The nutritional improvements that are needed will be evident from the Elm Farm soil analysis, and might include adjustment of pH (use natural lime, or dolomite if magnesium is also deficient), and phosphate (use rock phosphate – gafsa – on acid soils and redzlaag on alkali soils). If potash levels are indicated as low, there are two main means of correction: rock potash and kainit. Neither is ideal, because the former is considered to be almost too insoluble and the latter too soluble. The former is preferable, and with good biological activity, particularly in the region of the root, it should be effective and the mineral should become slowly available. The use of kainit is subject to approval by the Symbol committee and is therefore restricted only to special cases.

Other trace element amendments should be done as necessary. I have not mentioned nitrogen in this section as that is really in the domain of manure and leguminous crops in the rotation. Certain crops at certain times may need additional nitrogen, but the soil in general can produce its own if the biological activity is there and other factors are right.

Stocking Rate

The target for stocking rate is generally 1.2 livestock units per hectare, taken over the whole farm. In the ideal organic closed-system operation, this provides the best balance between livestock and crops and sufficient nutrient cycling capacity. For the small farmer this may be too much if, for economic reasons, the land down to arable or horticultural crops needs to be increased, and outside sources of fertility are available. Alternatively, with a more intensive livestock enterprise, using a greater proportion of bought in feed, this may be too low.

The organic farmer will aim for a greater proportion of a ruminant's nutritional requirements to be met from grazing than from concentrates. Choice of breed might be influenced by this principle. Grass growth is likely to start later in the spring, but carry on just as long, if not longer, in the autumn. This should be taken into account when formulating your fodder conservation policy.

Remember that it is preferable to have more than one type of livestock so that parasite and disease control, and pasture management, can be facilitated. Obviously the overall balance of the system must be reconciled with the realities of the situation in terms of finance, buildings and suitability. The small farmer may regard his livestock as secondary to his other enterprises, but their importance in an organic system should not be underestimated. Without them, or with fewer of them, other steps need to be taken to ensure that the absence of their nutrient-building and cycling contributions is covered.

Manure Handling

In many senses, manure is the central hub of the organic farm. What we do to our farmyard manure greatly affects the use to which we can put it. Straight from the barns and out on to the fields, it will either be rather strong with readily available nutrients, especially pig and poultry manure, or it will tend initially to rob the soil of nutrients if it contains more straw. The former might be useful for heavy feeding crops, but both types would benefit from what is often called semi-composting, to kill pathogens and weed seeds in the high temperatures that a heap will reach.

Well composted, the carbon to nitrogen ratio in the manure will have dropped from 20/30:1 down to about 10:1. The nitrogen will probably have actually increased by the action of free living nitrogen fixing organisms. However, potassium is likely to have been lost through leaching, and it is therefore important to be able to collect any run-off from stored or composting manure to save this rather mobile and problematic nutrient. The manure will have been largely converted into stable humus which will enrich the organic matter in the soil on a long-term basis. Light soil will therefore benefit best from compost, whilst heavier soil can take the semi-composted farmyard manure that is more normal. In Britain's temperate climate only certain circumstances require that you compost properly. It is not as critical as it is in the warmer countries, where biological activity is much more fierce and the organic matter must be as stable as possible.

If much vegetable waste is produced, it can be either circulated through livestock or composted. It is unlikely that a small farm will be producing slurry, but if this is the case, aeration of it is strongly recommended as this does for slurry what composting does for manure, cutting down both odour and nitrogen loss.

Bought-in manure should be composted as a matter of course, because the high temperatures will also help to break down any drugs or chemicals that may be present. Manure from intensive units, particularly the growing and fattening ones, should be analysed for heavy metals such as copper and zinc. Any other possible ingredients should be investigated too, since some do not break down and use of such manure might jeopardise the organic status of the farm.

Crop Rotations

This is one of the corner-stones of organic farming and thus of the conversion plan. Its function spreads through most aspects of the system. It must be designed to minimise weed, pest and disease problems, to optimise nutrient use and cycling and minimise nutrient loss, to build organic matter and soil structure, and, last but not least, to provide feed for livestock and returns for the farmer.

The conversion rotation will almost invariably begin with a ley, composed of a wide mixture of grasses, clovers and deep-rooting herbs. Table 3.1 gives a typical ley mixture, but is by no

A Typical Organic or Conversion Ley Mixture

Kg	Variety	Species
4	Frances	Perennial ryegrass
4	Merlinda	
4	Condesa	
2	Goliath	Timothy
2	Motim	
2	Farol	
4	Jesper	Cocksfoot
2	Salfat	Meadow Fescue
2	Dovey	Tall Fescue
0.5	Merviot	Red Clover
0.5	Grasslands Pawera	
0.5	Aberystwyth S148	Wild White Clover
0.5	Menna	White Clover
0.5	Alice	
0.5	Aran	
1	Chicory	Herb
2	Burnet	
1	Sheep's Parsley	
1	Ribgrass	
0.5	Yarrow	
Total	34.5 kg/ha	

Table 3.1

means the definitive one. This depends very much on your soil, climate, livestock and use. Under a ley the soil can regain its lost vitality, build up its structure and start its own nutrient cycling. It is a time when the worms can begin to multiply and life can slowly return. If a ley is not to be established, for instance when there is no livestock, an equivalent must be used. This will usually mean one or more green manure crops, such as rye and vetch over the winter, or mustard, lupins, red clover or some of many others over the rest of the year. White or subterranean clover can be used under glass, and the latter is also useful under soft fruit.

If the land is already in unfertilised pasture, its clover content and general health might suggest that it could go straight into cropping, but a short green manure might be advisable. A pasture without clover is not going to progress the conversion very far without being reseeded, and it would be unfair to make it try, for it needs the nitrogen fixation to help start the process, not to mention the benefits of the deep-rooting herbs that should also be there. The one thing to avoid is going straight from chemical arable into organic arable. It will not work, as I found to my cost when I first started.

The conversion ley can be undersown in the previous (chemical) crop, as can some of the green manures. After two, three or four years the rest of the rotation then follows. A heavy feeder should come next, but a weed-free period follows a ley, and this should be capitalised on if possible. Towards the middle of the rotation, there should be a crop with a bulky root system that stays in the ground, which the soil organisms can use to replenish their food supplies. Green manures should be included at any convenient opportunity to keep the soil covered, especially in winter, and to enhance nutrient cycling.

Manuring Plan

With a limited amount of manure, it is important to use it judiciously. The conversion ley will benefit from dressings, particularly of farmyard manure as extra feed for the multiplying soil life. If organic matter levels are low, this will be needed even more, or the developing micro-organisms will eat into what little is there and make the situation worse. Some say that

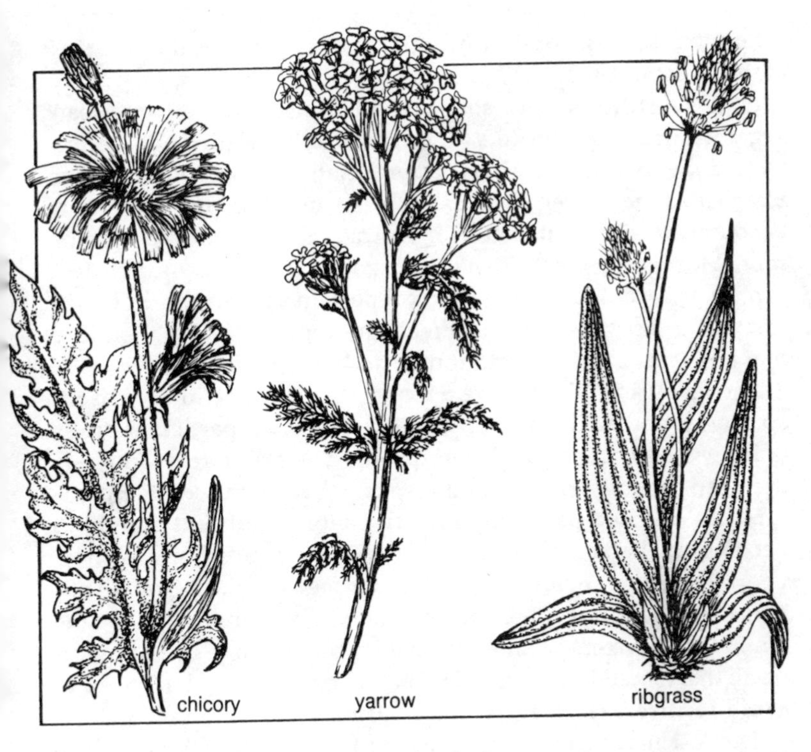

Deep-rooting herbs.

manure should be applied to green manure crops. This seems rather extravagant, unless it is to help establish the crop with a small amount of readily available nutrients, especially where the organic matter levels are so low that they need significant boosting.

Some crops are particularly demanding and some are less so. Some will even contribute to the nutrient status of the soil. These differences should be taken into account when considering the fertilising plan. Manure should be applied when it will have the greatest benefit. This means not only prior to a particular crop, or not after another one, but also at the right season. If spread in autumn or winter, much greater leaching of nitrogen and potash in the winter rains is likely, and so pollution too. Spring applications are best, or summer if appropriate to the cropping.

Other fertilisers, or 'soil amendments' as the American organic farmers prefer to say, might also be needed. These too should be put on at times appropriate to the rotation. For instance, lime, if required, should go on after potatoes, as it encourages scab, but before brassicas, as it discourages clubroot. Most of the rock dusts, that go on in smaller quantities, can be incorporated into the compost heap, where they will start to be activated. Their effects are long term and therefore one application in the rotation is probably sufficient.

The fertilisers which have a more immediate action are used, sparingly and with care, in conjunction with particular crops that need special growing conditions. The standard example is the need for early growth in the spring, before the soil temperature has warmed up enough for the natural nutrient supply to become available. This is a borderline practice because it is verging on the chemical approach of feeding the plant direct, hence the warning of 'sparingly and with care'. It is a time when the newcomer can easily panic, but when restraint might be all that is needed, coupled with the foresight to plan ahead to avoid recurrences in succeeding springs.

The last manuring practice is the use of foliar feeds such as seaweed and other 'tonics'. Their feed value is not necessarily their strong point, although seaweed is quite high in potash. The plant hormones, and a group of substances that they contain called alginates, are equally important. Their main use is as a tonic to enhance and invigorate plant growth and health. Seaweed is also a tonic to the soil, the alginates helping the soil structure among other things, and can be applied at any time during the growing season. Often seaweed is mixed with other ingredients, for instance in fungal control and with herbal sprays, to enhance their effects.

Other foliar feeds are more intended to provide nutrients, and are in the nature of a liquid manure for crops that need higher amounts of feeding than the soil would normally give. Comfrey preparations for glasshouse crops are one example. The liquid fraction collected from manure and compost heaps, and indeed silage clamps, can be diluted and used in this way too. The common factor between these types of liquid feed is that they are all high in potash, among other nutrients, which is a highly soluble and therefore potentially difficult nutrient. Those crops that need good quantities of potash are likely to benefit most from this practice.

Tillage

Your existing machinery will almost certainly need changing somewhat. The modern plough which can only go deep is no use on an organic farm. If one is to be used at all, it should be able to plough shallow, that is, no deeper than about 12–15cm, so that inversion is kept to a minimum. Many organic farmers do not use a plough, but prefer to employ other means of cultivating that have a less drastic effect on the environment of the soil life.

The powered implements that have come on the market in recent years show a marked improvement on the old Rotavator, which tends to smear the soil and produce an artificial tilth with its cutting blades. Avoid both effects in your newly organic soil. The Rotavator can however be used to chop up surface trash, going down only a few centimetres. The rotary or power harrow and the new spading machines are both useful tools. Their action is much more gentle, stirring and lifting, and the tilth is more natural and less susceptible to erosion and slumping as a result.

It is gratifying to see much of the current thinking in soil management today coming into line with what the organic movement has been advocating for many years. The popularity of subsoiling is a particular example. If you have not yet got a subsoiler, you should certainly be thinking of acquiring one now. Recognition of compaction is another, and horticultural methods like using the bed system, or ridges where appropriate, are very much in line with organic principles.

Unfortunately the increasing problem of soil erosion is less conducive to such easy solutions, and is unlikely to be reversed until more organic practices are incorporated into conventional agriculture. Where possible, cultivations and planting should be done with the contours, although this may present other problems with later inter-row work. Leaving the soil bare over winter is not good practice. Your tilth should be maintained by

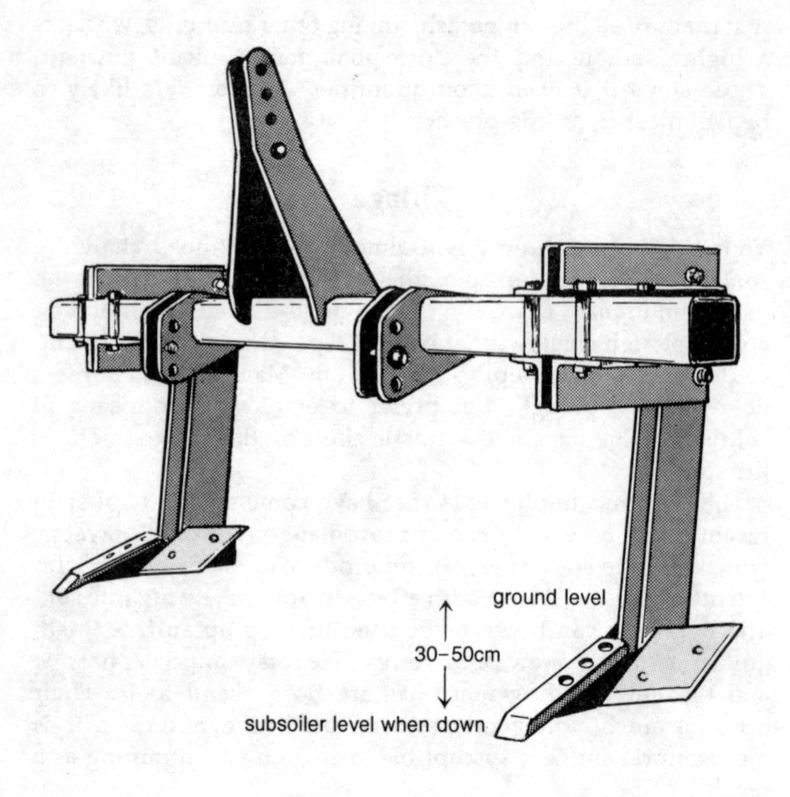

Subsoiler.

its biological health and its covering, except with particular soil types or in unusual circumstances, so in general the soil should be left covered as late as possible in the winter.

Weed Control

Weeds are one of the main problems for organic farmers and their control therefore extends right through all aspects of the organic system. Conventional control relies entirely on chemicals and the two opposite approaches highlight very clearly the difference between organic and conventional methods. Organic weed control starts with manuring policy and ensuring that weeds are not brought on to a field. It continues with seed-bed cultivations, the more normally recognised mechanical hoeing and the newer thermal methods. Then it is complemented overall by careful rotations, choice of varieties and the use of green manures.

Stop weed seeds coming on to the fields in the manure by ensuring that your composting methods create sufficiently high temperatures to kill them. Other ways by which weeds gain access to your fields, such as on implements, your boots and by air from headlands and hedgerows, should also be avoided. This last source can be prevented by keeping your hedges in trim. Troublesome weeds should not exist in a well-kept hedge, so this advice is thoroughly compatible with wildlife conservation.

Prepare your seed-beds about two weeks in advance of drilling the crop. This allows you to take a 'weed strike' of those seedlings in the top layers of the soil before the crop goes in, using chain harrows or something similar that works very shallow. This is called a 'bastard' or 'false' seed-bed and, if the weeds are expected to be really bad, it can be repeated a second time. In fact the action of drilling has the same effect as the preparation of seed-beds and can be delayed a further ten days after the first weed strike for that purpose.

For the slow germinating crops like carrots, thermal weed control can then be employed as a pre-emergence third line of defence. It uses very high temperatures acting on any spot for a very short time. It is not meant to burn the weed seedlings, but to raise their temperature just sufficiently to burst their cell membranes by expansion and dehydration of the cell contents, and to coagulate the cell protein; this happens above about 60 degrees Centigrade. Some crops can also take post-emergence flame weeding, for example maize before the leaves have unfurled and again after it has reached about 25cm, if the heat is aimed correctly.

Steerage hoes and other mechanical inter-row hoes all have their time and place for use. Timing is of the essence when dealing with weeds, and it could almost be said that the best time to kill weeds is before they come up. The larger they grow, the more difficult they are to kill, and before you know it they

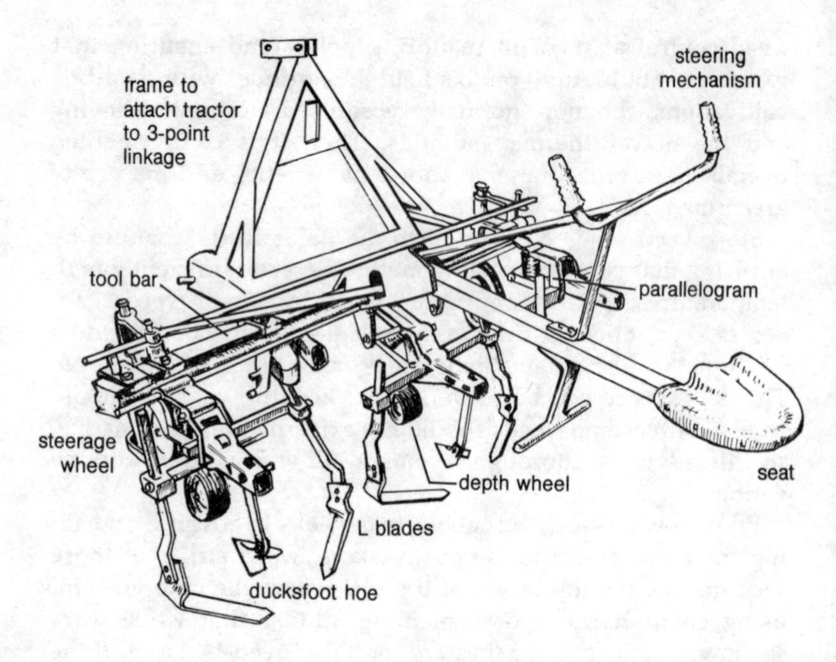

Steerage hoe.

have flowered and seeded and given you a problem for the next season. Even if weeds are hoed at the flowering stage, they can still go on to produce viable seeds, so it really is important to prevent them reaching that stage.

Inter-row weeding is one thing, but in-row weeding is another, for it cannot easily be done mechanically without damaging the crop. The last option is the hand hoe, but all the previous means are designed to minimise the need for this labour-intensive and therefore costly operation. Ridging taller crops, choice of the more vigorous varieties and manipulating plant spacings to obtain good smothering cover will all help to reduce the need for hand weeding further. Transplanting a crop, either bare-rooted or in modules, rather than sowing, can be another very important way of getting ahead of a serious weed problem.

Weed control must be looked at in the context of the long-term rotation, considering both the changing weed populations during the course of the rotation, and using the rotation itself as a weed-control measure. After the ley break, if there is one, the weed population is likely to be at its lowest, so the more sensitive crops should be slotted in accordingly. Smothering crops, or those used as 'cleaning' crops, can be alternated with them too. In other words the rotation is planned to accommodate an expected gradual weed build-up. It should be said that total weed control is highly unlikely, and may not actually be desirable. Weeds break up the monocrop and tend to attract beneficial insects like the natural predators and parasites. Aim for a balance so that weeds do not reach proportions that would affect yield and hence the economics of a crop.

Green manures can be used in the rotation to control weeds, among their other functions. They can keep the soil covered when otherwise it would be bare and, together with their regular cutting or grazing, can provide competition. They can also be undersown in an existing crop. In fact, green manures can almost be regarded as the organic farmer's cultivated weed substitutes. They enrich and help correct the soil and are used as cover for the soil, just as weeds do in the natural state.

Pest and Disease Control

Pests and diseases are often more a symptom of what has gone wrong or what is out of balance on an organic farm, rather than a hazard of the job. In deciding what crops to grow or what livestock to have, you should give due attention to whether they are suitable for your particular area and conditions, without unnecessary interference. You are trying to provide an optimum environment for growth. It follows that if you are on the edge of the ideal range for optimum growth, whether due to soil type or climate or another consideration, then you are potentially inviting problems.

It could be said that avoidance is the better part of control. Avoid putting sheep on wet ground that encourages foot rot and fluke, for instance. Similarly, if everyone around you is growing brassicas, it is probably wise not to, because there is likely to be a large population of cabbage root fly in the area. The organic farmer's aim of peak health and vitality for his stock

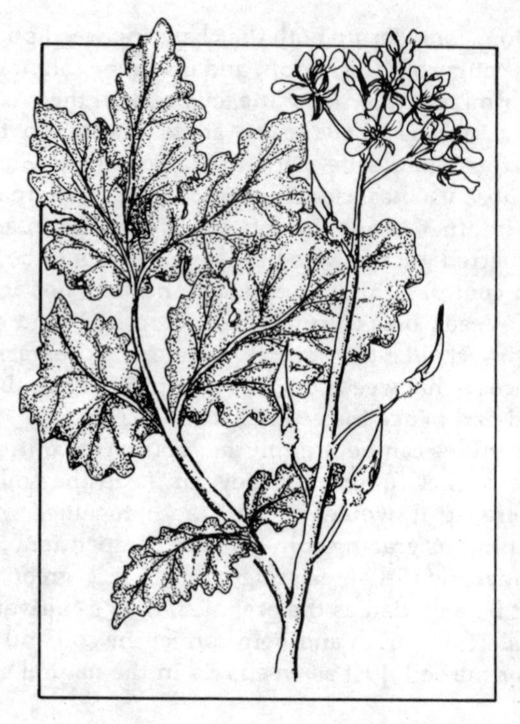

Mustard: a green manure crop and a preventive treatment for wireworm.

and crops is another general means of avoiding problems. This is better known as resistance, and extends to choosing resistant varieties or breeds, and ones that are generally more vigorous.

Once again the carefully planned rotation is important, of stock as well as of crops. This involves allowing a long enough period between similar crops to stop pests or diseases being carried over, and also perhaps incorporating crops or green manures that actually have a negative effect on certain organisms. For instance, two crops of mustard in quick succession will help to control wireworm if sown at double the normal rate of about 18kg/ha. Another green manure crop, *Tagetes Minuta*, will help to control eelworm (also the perennial weeds couch, ground elder and bindweed), but unfortunately it is not readily available in large quantities. It is probably best transplanted at about 30cm square in May and left to grow all summer.

On the smaller scale, companion planting is practical. This can mean combining two crops in the same space, and also including beneficial flowering plants or herbs in the crop rows, to attract or repel as the case may be. On the very large scale, the size of fields might need to be limited, to avoid having too large an area down to any one particular crop.

As with weeds, the aim is not to be completely rid of all signs of pests or diseases, but to ensure that none get to proportions that will cause economic loss. In a healthy organic crop you are likely to find the whole range of pests and diseases in small quantities, but only when something goes wrong will the balance be upset and a problem develop. Sometimes this may be caused by things outside your control, like weather conditions, but it can also be caused by over-feeding (of livestock too, particularly ruminants fed with too many concentrates), overstocking, and many other man-made factors.

Finally, specific control measures that are acceptable to organic farmers are available for some of the problems that might arise. Biological control, herbal preparations, homoeopathic remedies and insecticides of natural origin are all used. Natural insecticides break down harmlessly within days or hours, thus avoiding potential resistance build-up and long-term damage to the environment. Others such as antibiotics and copper fungicides are allowed, sparingly, as a last resort.

Machinery

The different cultivating equipment that will probably be needed in a new organic system, as well as inter-row and other weeding implements, have already been discussed. However, some other areas of machinery might need changing or modifying.

Obviously new crops or enterprises may require new equipment to service them and this will have to be acquired. The additional factors relevant to organics that should be considered at the same time really revolve around the impact of the new machinery on the soil, in terms of its structure and compaction. If there is a choice between several different machines that do the same job, choose the one that will be kindest to the soil.

With harvesting equipment ensure that it is both efficient and does minimum damage to the crop. The organic crop is a quality product and deserves to be treated as such. Mechanical damage in harvesting is not only unnecessary, it is costly. Remember that you might also need equipment for chopping up crop residues and green manures.

Manure handling comes in two stages: first composting, then spreading. The correct choice of a manure spreader should mean that, with a little ingenuity, the same machine will do both. One that off-loads from the rear, and is PTO (power take off) driven, can be used to make the compost windrow by standing still and discharging into a form behind it. A spreader which off-loads to one side is less easy to adapt, but may have other advantages.

When considering the level of mechanisation which you should employ on the farm, remember that the organic system will be more labour-intensive. If necessary, you may be able to

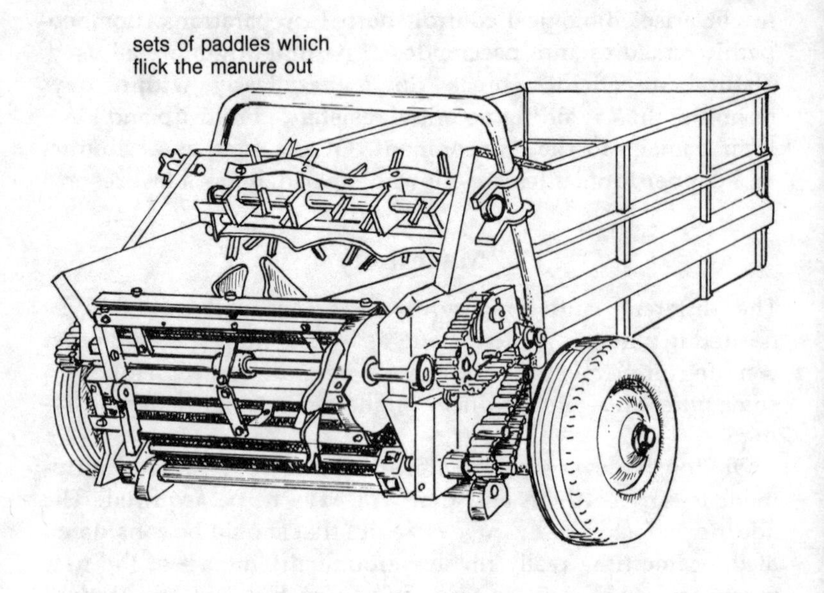

Rear off-loading manure spreader.

offset this, to a certain extent, by appropriate mechanisation. Alternatively you may feel that it is worth swinging the other way and becoming more labour-intensive still, if the opportunity is there, relying on casual labour, students and WWOOFers.

Labour

The importance of labour, and its quality, goes without saying in a system that relies on timing and attention to detail. It is therefore important to ensure that labour is not short for the critical operations of the season, such as hoeing, harvesting and planting; also that critical operations do not clash in their labour requirements, such as lambing with planting, or hoeing one crop with harvesting another. The best way to be sure of all this is to draw up labour profiles for each enterprise or crop through the year, put them together and juggle them around until you are satisfied that you have a system that will work. At least get to know where the peaks are and plan accordingly.

One of the hidden principles of organic agriculture involves an appreciation of the quality of life of all human beings. We should respect our livestock, and we should also respect our fellow humans, especially those who are working with or for us. An equitable relationship between those that live and work on the land is part of that principle. Another thought to bear in mind is that many of those who seek work on organic farms do so to learn. These human aspects of labour should not be forgotten.

Marketing

One of the great advantages of the organic produce that you will be selling is its quality. Perhaps its external quality might not quite match up to the best conventional standards in some products, but internally it should be in every way superior. Flavour, nutritional quality, lower residues and storage life – all these are advantages in your favour, for which an increasing number of the general public are prepared to pay a premium. However, in order to reap this benefit, you have to do two very important things: first you must seek out and establish a market, and then you must market your produce effectively.

Markets and marketing will be discussed more fully in chapter 4, but a few general remarks are appropriate to the conversion period.

It is pertinent to reflect on what will be produced from the fields under conversion. Except in intensive farming, the conversion period should be spent in the conversion ley, and this means that only livestock or livestock products are going to be available for sale. Assuming that the proportion of the farm being converted at any one time is going to be relatively small, it is highly unlikely that different livestock will be kept on the different sections. This means that their organic purity cannot be verified. You will probably have to accept and use conventional markets for your conversion produce in the main. However, there is increasing interest in free-range, hormone-free meat for which a premium is available, its size depending on the type of meat. This may provide some help during these intervening years.

Growing intensively, you will have conversion produce for sale after one or more green manure crops. A 'conversion grade' has just been developed, which will be implemented in the 1987 season. What premiums it will attract and how it will work are unclear at present, but it is designed to assist the converting farm and its finances through this most difficult time, and also to help regulate this period to the benefit of the organic scene generally.

An alternative, but one that I do not recommend, would be to pretend for a year or two that you have 'conservation grade' produce and join the Guild of Conservation Farmers. You will easily qualify under their 'low input' standards, but the premiums are not nearly as great as the organic ones. It must also be very strongly emphasised that the conservation grade standards are not up to those required in the conversion period. The temptation to stay with conservation grade will get you no nearer to organic than if you had stayed with conventional practices. Unfortunately, conversion and conservation are two easily confused words. Be warned: they are not the same.

You should remember that the premiums paid for fully organic produce are not necessarily available for conversion produce. It rather depends on the extent to which demand outstrips the organic supply. This obviously cannot be easily foreseen and should not be relied upon, but the premium can nevertheless be substantial.

When considering your farm's progress to its full organic status you obviously need to plan for the proper marketing of your organic crops, even though you may not be using these avenues at first. You should be doing some serious research and making the right contacts in preparation for this time.

Yield Estimates

The administrative and economic aspects of your conversion are also important. What will be the implications of all these new policies for you and your farm?

The yields that you can expect from your conversion leys will be considerably reduced compared with what you have been used to, and also compared with what those leys will produce next time around in the rotation. However it has been said that if you apply up to 200kg of nitrogen per hectare on your conventional pasture, you might as well throw it down the drain. A decent clover-rich sward will produce the same growth absolutely free. There was even some research done by the Grassland Research Institute a few years ago which suggested that the yield from grass receiving nitrogen up to 400kg/ha could be matched by a clover-rich pasture receiving nothing. Yield in this instance was taken as the growth rate of the livestock on it, and the reason given was that even though the fertilised grass produced more bulk, the clover was some 25 per cent more palatable.

Other fully organic crops are likely to be some 20 to 30 per cent lower in yield than their conventional counterparts. During conversion, they may well be lower still by the same amount again. Variations on these figures can obviously be enormous, due to the vagaries of the weather or pests and diseases. In crops in particular the need for special row spacings and weed control influences yield. However, variations due to the weather are likely to affect conventional yields more than the organic, because the soil environment is that much more stable in the organic system.

The difference in yield that you will experience on your own farm will depend as much on your present level of intensity, the health of your soil and the quantity of fertility-building manures at your disposal as on the other factors mentioned above.

Financial Implications

It is now time to sit down and work out the costs and returns of the conversion process. Because it is likely to be occurring on only a small part of the farm at any one time, this is not an easy task. One way is to take the intended progression of the rotation and look at it on a per hectare basis, whilst recognising that the influence of the rest of the farm on the rotation will change as more land becomes organic. This will show up when the livestock can be considered organic and as your own experience develops.

For the conversion ley, variable costs will include seed, extra cultivations and soil amendments, but not fertilisers and sprays. For other parts of the rotation, variable costs will also include manure spreading, extra weed-control measures and any casual labour that might be needed. Fixed cost changes will depend on how much the new system requires different equipment, a different labour structure and modification to buildings.

Returns will depend on the lower yield, mistakes, the development of new crops and enterprises and the organic premiums for them. In addition, the livestock enterprises may well require smaller veterinary bills and lower concentrate costs, but possibly more home-grown feed or bought-in organic feed and lower stocking rates. As the farm settles down, the possibility of adding value to what is produced by on-farm processing might become feasible.

It might also be pertinent to consider any other ways that you could use to bring in additional cash at this time. Offering bed and breakfast accommodation is a classic sideline for farms, if you have sufficient room. A part-time job is another possibility, but you always run the risk of taking valuable hours away from more vital work on the farm. Some jobs do not clash as much, such as relief milking, milk recording or evening work in pubs, for instance, and they would therefore be better. Others such as agricultural contracting might be appropriate if it uses your existing machinery, but only if it does not get too demanding.

Alternatively, you could do some work for a neighbouring farmer, possibly in exchange for the use of some of his equipment.

Remember that a true picture of the farm's finances is not really possible unless it is considered as a whole. Individual crops or enterprises must always be seen in that context, for their contribution might be in other ways than financial. The conversion process is a stepping-stone to the new system and its costs must be considered in that way too.

Timetable

The two most important factors that should have an influence on the timetable for the conversion of your farm are firstly the financial implications, and secondly the level of your experience. You must obviously proceed only as fast as the farm's profitability can cope with, as this initial period is likely to be the most critical from an economic point of view. The heavier your debt burden or the lower your profitability, the longer the conversion should be planned to take, so that the pressure on any one year's income can be reduced.

Organic farming requires an additional set of skills and knowledge, as should be clear by now. Whilst these are being acquired in the first years, the risk of mistakes and misjudgements will be much higher. It is therefore best to restrict this danger to as small an area as possible. Your first land to be converted should preferably be confined to just a few hectares or a field or two, depending on the size of the farm.

Do not convert more for at least another year or even two. This will give you time to gain experience with the conversion ley and the new situation and new thinking that comes with it, before you commit more land to the process. The first fields act as your own trial grounds and any problems that arise can be dealt with at that level where the impact will still be small. Plans can then be altered accordingly for the rest of the land that will be following.

As your experience grows and as your system is perfected, the pace of the whole farm's conversion can be speeded up. By this time you will probably be impatient to get on with it so that you can banish chemical fertilisers and sprays from your farm as soon as possible.

4
Input and Output

Before you contemplate full conversion, you will need more information about means of providing fertility in an organic system and the overall picture of nutrient cycling – a term which has been used but not really defined. You will also need more extensive knowledge of the product and its marketing – a subject of great importance if the organic premiums, on which we depend at the moment, are to be realised.

MAINTAINING FERTILITY

The ideal organic farm is as closed a system as possible. The livestock should be fed almost entirely from the farm. The crops should be fed by the soil, and the soil fed by the animal and plant by-products of the farm. There are several reasons why this is possible and can be sustained whilst providing surplus produce for sale; firstly, photosynthesis is a building process, taking sunshine and carbon dioxide from the air and converting them into energy-containing carbon compounds for the plant. Secondly, the nitrogen fixing bacteria, which are both free-living and associated with the leguminous crops in the rotation, bring nitrogen into the system from the air. Thirdly, the soil life works on the inorganic fractions of the soil and makes available the otherwise unavailable minerals locked up in their structures. This may be considered to be a form of mining the soil, but the quantities of minerals locked up are thousands of times more than those taken out of the system by selling produce off the farm, and other inputs, like rainwater, may well put back a sizeable proportion of them.

In the small organic farm, such idealism may be hard to aspire to. The financial situation probably requires that the ratio of livestock to crops is lower than the desired 1.2 livestock

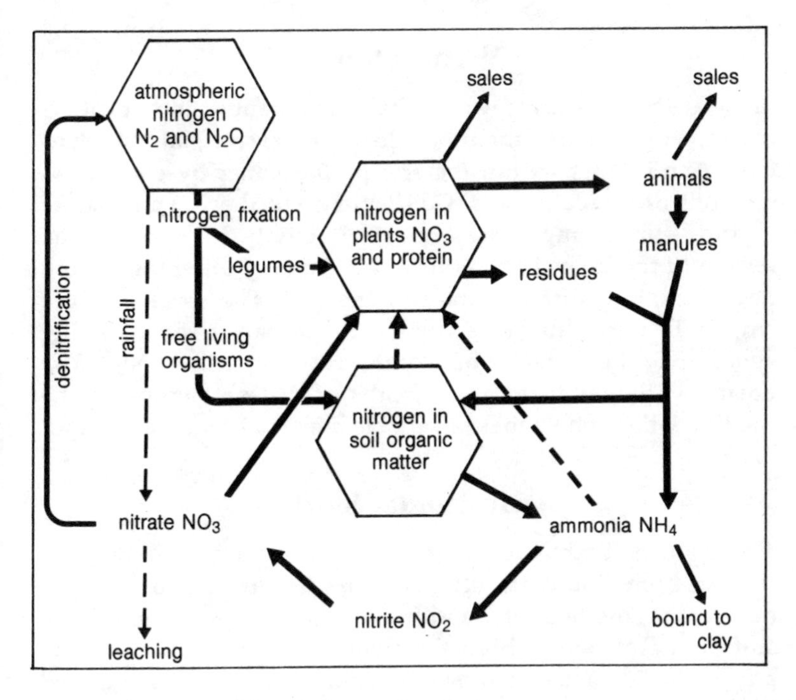

The nutrient cycles of nitrogen.

unit per hectare, taken over the whole farm, and that arable or, more likely, horticultural enterprises will predominate. Livestock enterprises may rely on bought-in rather than home-grown feed to sustain them, and it may be that the farm is without livestock altogether.

CONTROLLING NUTRIENTS

The problem that confronts a small farm is not only one of quality but also of quantity; not only knowing how to build up fertility, but knowing how much is needed and of what. Knowledge of the correct quantities of the various substances should throw further light on the formulation of the rotation, the manure handling and the livestock.

Vertical Shift

There are broadly three ways of handling manure and nutrients in a system. The first might be called vertical shift, and involves moving nutrients within the soil profile either by cultivation methods or by plant roots. Cultivations stir the soil and aerate it, thus stimulating more biological activity to work on the release of the unavailable nutrients. Plants of different rooting depths exploit different areas of the soil. The green manure crops will move nutrients from their perhaps deeper-rooting zones up to the surface where other crops can utilise them. The greater variation in rooting depths of different crops in the rotation will in effect make available more nutrients in the soil.

Horizontal Shift

The legumes, both crop and green manure plants, will perform the same function with nitrogen. But they are also involved in the second method of handling nutrients, called horizontal shift. In simple terms this is the manure aspect, where nutrients are produced or taken from one area of the farm, are cycled through livestock or the compost heap, or both, and then moved to another area of the farm. Included in this are other forms of stock feed. Straw or other plant by-products are the agents for moving that food of the soil life, carbon, about the farm.

Clover-rich pastures are therefore seen to have two functions from a nutrient point of view; that of resting and building up the fertility of their own soil, and also that of providing nutrients for other cropping parts of the farm, initially in the form of hay or silage. The permanent pasture of a farm will lessen the pressure on the leys to produce nutrients for horizontal shift, but will increase the pressure on the arable land as the rotation will have to be that much more intensive.

Buying In

It should now be clear why the careful conservation and utilisation of all these potential sources of nutrients is so important. The balance between deficit and gain is very fine. To buy nutrients in from outside, the third method, is costly and is not really organic in its true holistic sense. However, all farmers, and small farmers in particular, are going to use this method to some extent on some occasions and with some particular nutrients. Remember too that buying in feed for livestock, or straw, is another form of buying in and introducing nutrients into the system. Similarly, selling hay or manure is another way of moving nutrients off the farm, of selling off its fertility, and is the reason why tenancy agreements of old used to ban these two practices. Indeed, it was considered such bad farming that the tenant could even lose his tenancy if he was found doing this.

BUILDING FERTILITY

By looking at the constituents and values of the various forms of fertility-building materials and methods, a nutrient budget can be constructed. These will be very general figures, to be taken as rough guidelines and indications of what is required. To put this in context, the first material to investigate is the soil itself.

Average Nutrient Levels in an Arable Organic Soil

	N[1]	P[2]	K[3]
Average percentage in topsoil	0.25	0.07	2.5
Kg in 1ha of topsoil	6,270	1,750	62,720

1: Nitrogen
2: Phosphorus
3: Potassium

Table 4.1

The table above shows quantities of the main nutrients in an average soil. Compare these to the amount taken out by cropping (Table 4.2) and you might almost be forgiven for thinking that the soil's reserves are relatively inexhaustible. Two other factors bear on those figures, however. First, they include the substantial amounts of nutrients locked up in the mineral and humus parts of the soil, which are, to a certain extent, in dynamic balance with the much lesser amounts that are actually available. Second, their availability depends on the soil life, (which itself accounts for another sizeable portion of nutrients), and which requires feeding to keep active. The system therefore needs continual priming if it is to operate at the level we expect.

Average Nutrient Levels in some Common Crops

	%N	kgN	%P	kgP	%K	kgK
Wheat	2		0.3		0.44	
6t/ha		120		18		26.4
Potatoes	0.32		0.06		0.4	
30t/ha		96		18		120
Onions	0.24		0.036		0.157	
30t/ha		72		11		47
Swedes	0.2		0.04		0.25	
40t/ha		80		16		100
Carrots	0.18		0.036		0.341	
30t/ha		54		11		102
Cauliflowers	0.43		0.056		0.259	
30t/ha		129		16.8		78
Courgettes	0.175		0.029		0.202	
33t/ha		58		10		67
Lettuce	0.19		0.026		0.265	
10t/ha		19		2.6		26.5
Cabbage	0.25		0.03		0.5	
40t/ha		100		12		200
Tomatoes[1]	0.175		0.027		0.244	
250t/ha		438		67.5		610
Peas	1		0.02		0.316	
6t/ha		60		1.2		19
Grass/clover	0.48		0.065		0.45	
45t/ha		216		29.3		202

1: Long season, heated crop

Table 4.2

Having due regard for the necessarily highly approximate nature of the figures in this table (despite the number of decimal places), the large variation between different crop types is nevertheless clear. Note how minimal is the draining of the soil's reserves when a leguminous crop like peas is grown, especially when its ample haulm is left behind or recirculated through livestock. Furthermore, you can see why a horticultural crop like cabbage is in a different league to the arable crops like wheat, and a protected crop like tomatoes is in a league all of its own. Somewhere in the middle come the root crops, with potatoes topping the bill. Pasture is also seen to be a highly demanding crop, but a reasonable proportion is returned direct when it is grazed and nitrogen fixation is still adding underground almost as much as it is providing above ground.

Returning Nutrients to the System

Now let us look at the different ways of putting nutrients back into the system. In the case of nitrogen, the leguminous crops are the most important. Their contributions are listed in the

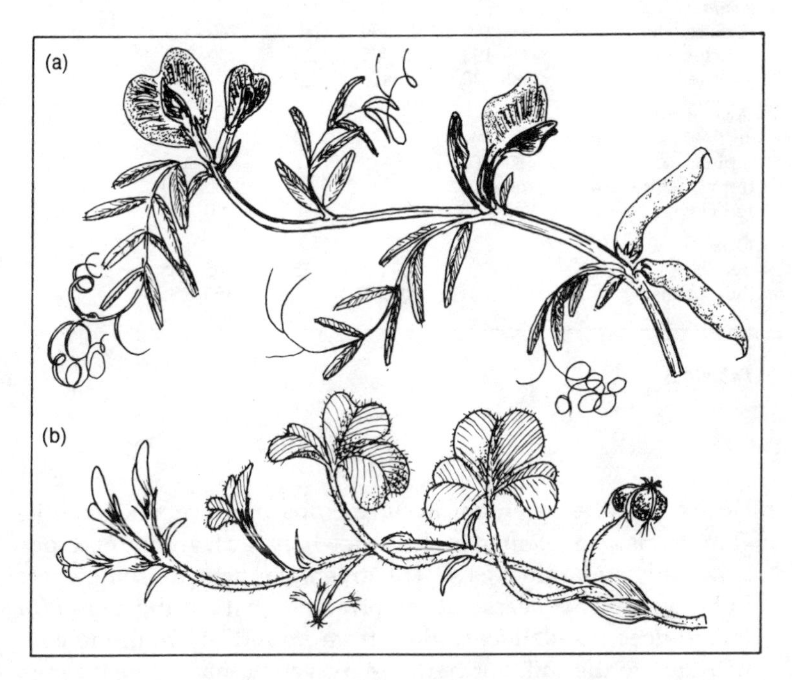

(a) common vetch (b) subterranean clover.

table below. The figures are mostly taken from European sources, but are generally applicable in Great Britain. The variation between the highest and lowest amounts can be large, depending on the growing conditions, but the value of their nitrogen fixation is obvious. Lucerne seems to give the best results, but it is best suited to drier conditions, as found in Europe, and so is of less importance in Britain. It is also a perennial crop that takes at least the first year to establish itself, and its upright habit means that it is best not grazed. By contrast, the clovers are the outstanding plants for British conditions, as the predominance of pasture in this country suggests.

Nitrogen Fixation in Various Legumes: kg/ha/year

	gain in soil	sales/feed	other loss	total fixation
Pasture:				
grass/clover	60–180	150–250	30	240–450
red clover	80–100	125–325	25	230–450
lucerne	100–120	180–400	20	300–540
Green manures:				
vetches	50–80		10	60–90
lupins	50–100		10	60–110
trefoil, white clovers	60–100		10	70–110
red clovers	80–120		10	90–130
Crops:				
peas	30–50	60–180	15	105–245
field beans	50–70	90–200	15	155–285

Table 4.3

Notice that the perennial pasture crops produce a substantial gain that is also retained in the soil, where the leguminous crops leave only a marginal gain. The green manures are somewhere in between. This bears out the principle that cutting a crop for forage does not detract unduly from its fertility-building contribution to the soil. Furthermore, recycling that forage through livestock and then back to the soil can actually enhance its value. This is one aspect of nutrient cycling.

There are a few other sources of nitrogen input that can be taken into account, but they are small and can be considered to replace the small losses that inevitably occur in even the most carefully worked out system. The first is from the soil's own free-living nitrogen-fixing organisms which produce up to 15kg/ha/year. The second is from the rain which can contain up to 20kg/ha/year.

Of the other nutrients, there is usually little need to worry about phosphorus, except on some upland and acid or alkaline soils where it may be deficient or, more likely, locked up. Its level in the soil is very stable because of its insoluble nature and it is only by the action of the soil life that it is made available. Where land has been in good cultivation for many years and phosphate fertilisers have been consistently applied, the levels will probably be high because only about a quarter of it will ever have been used by the crops. The rest stays in the soil, building up all the time and waiting for good biological activity to come and release it. Between that and the phosphates in the returned manure and compost, you should have enough. The organic soil analysis will tell you all you need to know if the situation is any different.

Potassium is a bit more tricky. It is a very soluble ion and is required in quite large amounts by plants. Unfortunately the large quantities in the soil are mostly locked up, for they are an integral part of the chemical structure of clay particles. Only a tiny proportion (about 1 per cent) of the potassium is actually exchangeable, although there is a dynamic equilibrium between the soluble and insoluble parts that keeps the exchangeable part topped up.

Fortunately, potassium is not an important nutrient for livestock. They obviously need some, but it is constantly mobile through their system and is excreted in their urine. It is therefore important to have very efficient collection and storage facilities for the liquid fraction of manure, otherwise the potassium drain can be considerable. It was commonplace in the farms of old to have an underground storage tank where the liquid manure was collected and held. It is a practice well worth reviving.

Average Percentages of Nitrogen, Phosphorus and Potassium in some Common Manures

	t/year	water	N	P[1]	K[2]
Solid:					
dairy cows	11.6	80	0.3	0.07	0.4
beef		80	0.2	0.13	0.3
sheep	2.3	80	0.3	0.11	0.34
pigs (sow)	4.3	80	0.4	0.2	0.24
Hens:					
layers (pit)	53kg	75	2.5	0.3	1.5
broilers (litter)	0.43kg[3]	75	1.5	0.5	1.2
Liquid, urine (cow)	4.9	95	1	0.01	1.5
Slurry (cow)	16.5	90	0.3	0.05	0.73
Farmyard manure		80	0.3	0.17	0.5
Compost		65	0.5	0.3	0.6
Digested sewage sludge		50	1.4	0.4	0.1

1: multiply by 2.29 to obtain P_2O_5, phosphate.
2: multiply by 1.2 to obtain K_2O, potash.
3: quantity given as kg/week

Table 4.4

The figures in table 4.4 must only be regarded as very approximate. Whilst researching for all these tables, but this one in particular, I found huge differences between the figures from different sources. The composition of the manures can vary enormously according to the time of year, the feeding regime and the moisture content. Some conclusions can be drawn, however: dairy cows and laying hens, the breeding and producing stock, excrete more nitrogen and less phosphorus than their equivalent fattening stock. This reflects the different demands that their occupations make on their systems, and to some extent the different feeding regimes that are employed. The amount of potassium in urine is plain to see, as it is to a lesser extent in slurry. Compost comes out pretty well, being slightly more concentrated than farmyard manure, and its nutrients are in a more stable form. Remember too that sewage sludge contains substances such as heavy metals which may make it less desirable.

One other constituent to bring into this picture is the bedding, usually straw. It must perform the additional function, besides providing clean and dry quarters for the livestock to use, of balancing the nutrients in the dung. Too little straw, the usual complaint, is not an economy as it will only result in excessive wastage of nutrients from the resultant heap, not to mention the unhygienic and inhospitable conditions which the animals will be subjected to. Something in the order of 6–10kg of straw per livestock unit per day is considered suitable, depending on the type of livestock and their level of feeding. The aim is to achieve a carbon to nitrogen ratio of about 25:1.

If you are buying in nutrients from outside, manure from neighbouring farms can be assessed by using the table above. Do remember, however, that some manures, particularly from intensive fattening units, are likely to contain various unwanted constituents that may jeopardise your organic status. An analysis should be made of them, to check especially for heavy metal content, and they must certainly be properly composted before use.

COMPOST

There are two stages to the composting process. In the first, the main requirement is for the heap to heat up to about 75 degrees Centigrade. This will kill most weed seeds and pathogens, will neutralise any antibiotics, and will cause the initial rapid decomposition of the raw materials. The heap at this point will be well-rotted manure and this is the most common form of manure found and used on British farms. The second stage is the maturation phase where nutrients are further stabilised and humus is formed, producing proper compost, some of which will still be detectable ten years after application to the soil.

Compost making is an aerobic process because the fungi and bacteria need air to live and do their work. This consideration therefore determines the conditions required in the heap. It must not be a compact, waterlogged and acidic mass, but needs to be of fairly open, even texture, damp and of neutral pH. Furthermore, the outside of the heap will not reach the necessary temperatures unless the heap is extremely well made and

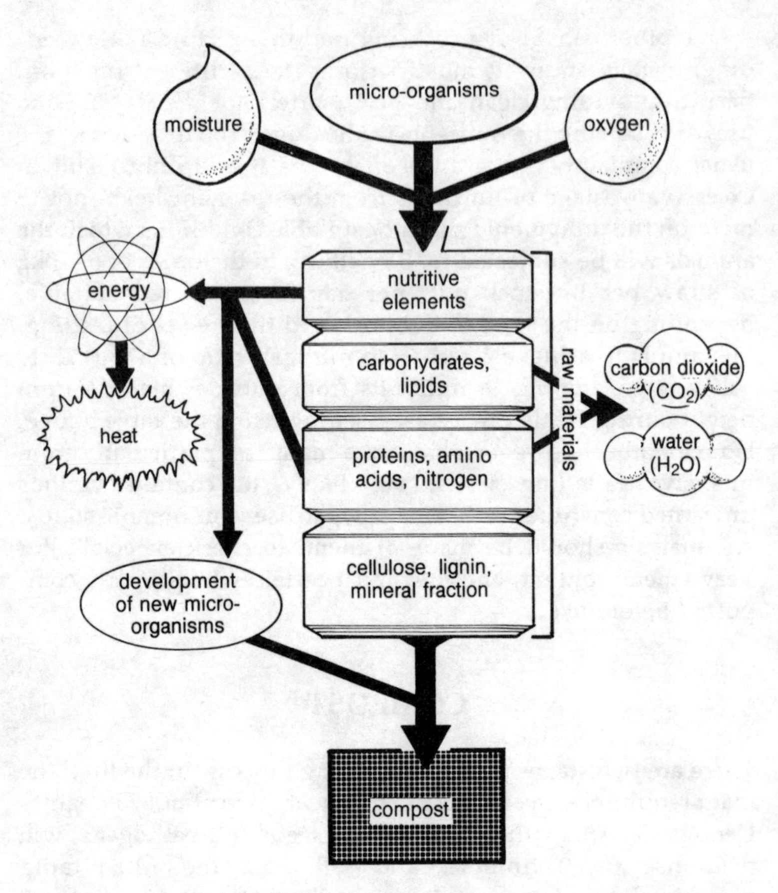

The chemistry of the composting process.

insulated. The whole thing will therefore almost certainly need turning and mixing half-way through the process, to ensure that the outside gets an equal opportunity to heat up.

Constructing the Compost Heap

This is what you should do. First, make sure that the manure is of the right consistency, damp but not so wet that any water can be squeezed out, and limed if it is acidic. Second, make the heap. An ideal heap needs to have a large surface area, balancing maximum access of air with sufficient size to allow good heating up. The best shape on a farm scale is probably a line heap some three metres wide and no more than two metres high, flattened on the top. You can bore holes into the centre from the sides to allow better access of air into it. It can be as long as you like and would be ideal along a farm track, for instance. The best machine to make something like this would be a rear delivery manure spreader, which can offload into a form whilst stationary. It can then be moved along as the heap is created behind it.

Other ingredients besides manure, such as vegetable residues, can all be added as well, though they may need shredding or chopping up if they are too large and woody. The alternative is to incorporate them into the livestock bedding as they are produced, or better still, process them through the livestock as feed. Rock dusts and any other ingredients are best applied to the bedding as it builds up, or sprinkled on each spreader load before offloading.

Turn the heap after a few weeks when the temperature inside it starts to drop. You can either repeat the building process or

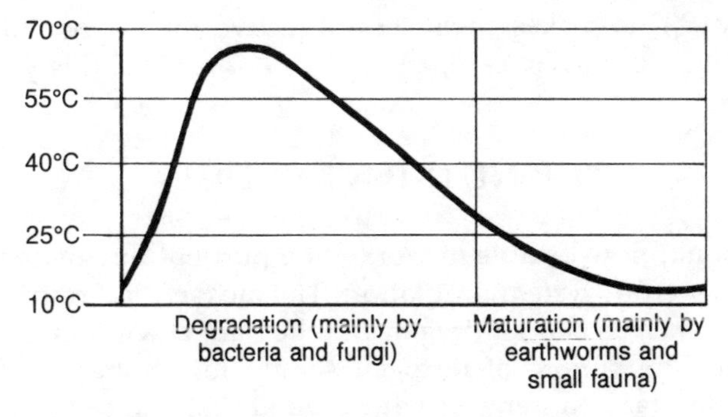

Temperature variation during the composting process.

just move it over with a fore-end loader, for the mixing does not need to be too exact. If you are going to leave it for any length of time before spreading, it is best to cover it with a layer of straw to protect it from the extremes of the weather. Sunshine will dry it out and rain will cause both leaching and waterlogging. A good heap will be ready after about two months in the summer, but it may take up to six months in the winter.

This process makes good and proper compost. Less than the ideal, which basically means only stage one, is perfectly acceptable for most purposes. There are certain exceptions, however. The first and most important is for bought-in manures, particularly from intensive units in which all sorts of additives are fed to the animals. These must have the maximum amount of time and opportunity to be broken down so that they do not get as far as your soil. Even then, some, such as chemical hormones, will survive. The second exception is for very light soils where the stable organic matter levels need to be built up to improve structure and water-holding capacity. The third is similar. On soils that are very low in organic matter, compost is the best way of introducing more. Lastly, as one goes further south into hotter climes, the need for the most stable form of organic matter becomes ever more critical.

Like manures, the constituents of organic fertilisers can vary enormously depending on the precise source of the ingredients. One thing to watch out for with some of the general fertilisers is that they may have had soluble nutrients added to them in order to make their analysis more attractive or even just more constant. This makes them less attractive for us, not to say prohibited, but some companies have not realised this yet.

THE NUTRIENT BUDGET

You should now be able to work out a nutrient budget for any particular crop, system or rotation. The most critical nutrient is usually taken as being nitrogen and the budget will concentrate on that. The supply of nitrogen should ideally be met from within the farm system, or with minimal assistance from outside sources. It is the job of the nutrient budget to help identify correct rotation and manuring practice in this context.

Analysis of some Common Organic Fertilisers

	N[1]	P_2O_5[2]	K_2O[3]	Ca[4]	Mg[5]
Trade:					
Humber	4.5	2	0.5		
Zevel	2.5–3.5	2.4	2.8		
General:					
Dried blood	13				
Hoof and horn	14	2			
Wool shoddy	5–14	2	1		
Fish blood and bone	6	7	6		
Meat and bone	7	14			
Bone meal	4	20			60
Rocks:					
Phosphate, Gafsa		28			
Redzlaag		32			
Potash			11		
Feldspar potash			10		
Seaweed:					
Calcified		0.35	0.2	45	2–6
Meal	1	0.3	2.5		
Liquid	0.5	0.25	2		
Comfrey liquid (95% water)	0.12	0.05	0.86		

1: Nitrogen
2: Phosphate (phosphorus pentoxide)
3: Potash (potassium oxide)
4: Calcium
5: Magnesium

Table 4.5

As we have seen, potash usage should also be carefully considered. The leafy crops, like cabbage or hay can mean the export of considerable quantities of potash, so when choosing your rotation and crop management this is the next most important nutrient to include in your calculations. There is no means of fixing potassium, but you can ensure that your chosen rotation does not use excessive amounts of it and that your manuring practice is geared to the demands on it.

Two examples will help to clarify what you have to do. However, every situation is different. These rotations should not therefore be taken as gospel, but merely as indications of general principles.

Nutrient Budget for Nitrogen and Potassium in kg/ha

Rotation	N fixed	N in	N sales	K in	K sales
Farmyard manure 50t/ha		150		250	
Potatoes			100		120
Broad beans	200		150		100
Green manure[1]					
Farmyard manure 25t/ha		75		125	
Spring cabbage			100		200
Spinach			100		100
Compost 20t/ha		100		120	
Onions			70		50
Green manure[1]					
Totals	200	325	520	495	570

N balance is 200 + 325 − 520 = +5
K balance is 495 − 570 = −75

1: The green manures do not need to be legumes, but the first one could be trefoil and the second rye/vetch. Rye would do for both.

Table 4.6

Rotation	N fixed	N in	N sales	K in	K sales
3-year *Ley* (grass/clover)	700		500		250
2×slurry 10t/ha		60		145	
Winter wheat			120		25
Green manure (vetch/rye)					
Farmyard manure 25t/ha		75		125	
Potatoes			100		120
Carrots			50		100
Farmyard manure 15t/ha		45		75	
Spring barley (ley undersown)			100		30
Totals	700	180	870	345	525

N balance is 700 + 180 − 870 = +10
K balance is 345 − 525 = −180

Table 4.7

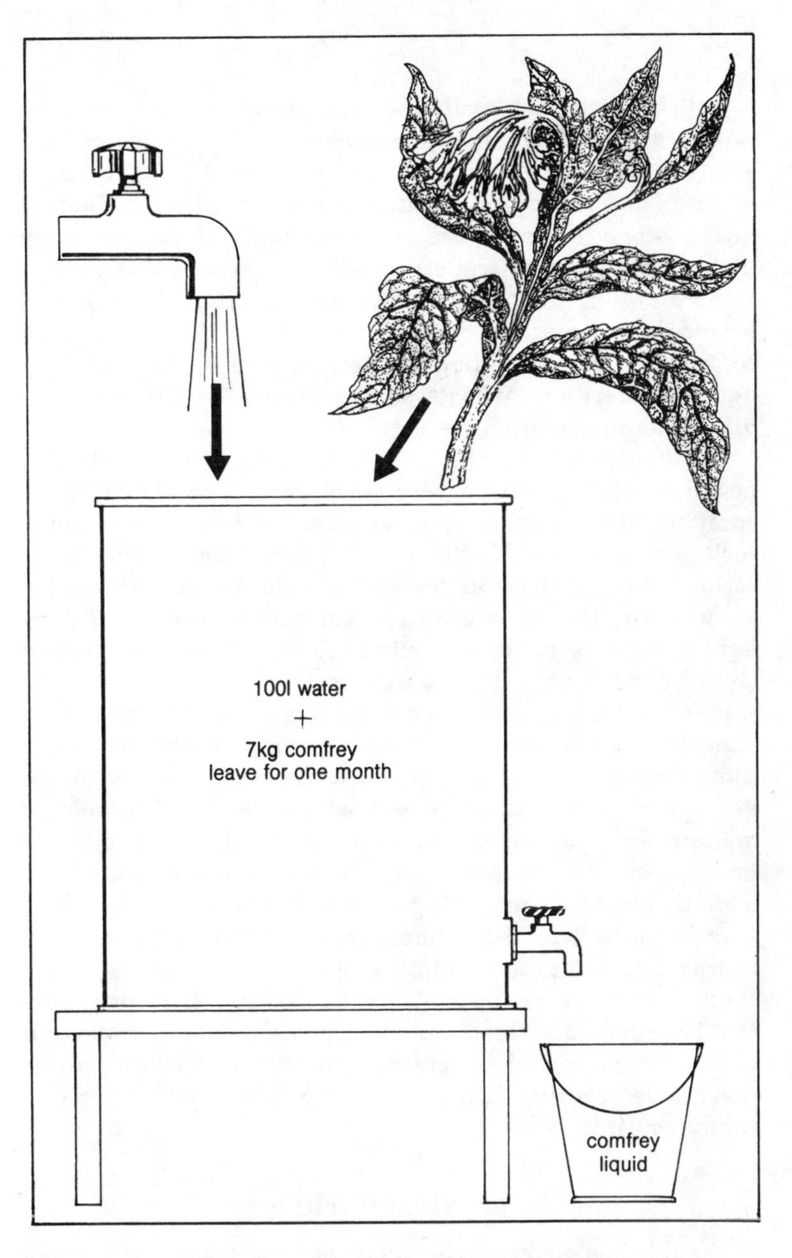

Comfrey liquid production: comfrey should be planted a metre apart, each way, in permanent beds, and can be cut every six weeks through the growing season. It is a plant and animal food.

In both examples the small losses of nitrogen are assumed to be cancelled out by the other miscellaneous gains. But notice that, where possible, the ground is left covered over winter, either by green manure or crop to reduce this potential drain. There is further scope for this in the second example, if the harvesting times of the potatoes and carrots allow it. Without taking into account any fixation by the green manure crops, the nitrogen balance is just positive in both rotations, which allows for a continuing build-up of fertility. Remember of course that these figures are very approximate, so that the real-life variations may differ enormously from these averages.

On the other hand the potassium balance is considerably negative. This can, to a very small extent, be alleviated by spraying all crops with liquid seaweed, and perhaps by using rock potash. However, although the potassium balance looks serious, the reserves in the soil should be able to supply enough. You should only need to worry if you have particularly light soil or its biological activity is low. Your soil analysis should show up any cause for concern.

It is interesting to note how moderate are the amounts of manure or other fertilisers that are actually needed to replace the nutrients which the crops have taken out. I know of one very good conventional grower who spreads 100 tonnes of manure every year on each hectare of his land. Thank goodness he does not use fertiliser as well! But it comes as no surprise to learn that he uses the whole gamut of chemical pesticides. I am sure he has to, with the lushness of growth that that quantity of nutrients would produce in his crops.

Other rotations can be worked out following all the principles that have been discussed, but these two give an idea of the sort of intensive and more extensive types that one might find on the smaller organic farms where vegetables have to play a prominent role.

Nutrient Cycling

It should now be clear what is meant by the term 'nutrient cycling': it means keeping the nutrients in a system active, keeping them moving. We encourage this seemingly pointless occupation because it fulfils three roles that are of prime importance to productive and sustainable organic agriculture.

1. Keeping nutrients in circulation in the system ensures that they are in use and cannot be lost by leaching away. Keeping the soil covered with green manures, if not with crops, and careful handling of manures are all part of this, as are the deep-rooting plants that can reach those nutrients that have almost been lost.
2. It involves using certain plants like the leguminous crops that can build more nutrients into the system at the same time. Circulation of nutrients through livestock and through the compost heap also contributes to this building process.
3. These very actions all ensure that the soil life is actively involved in the system. Its role is crucial to the working of the organic soil and nutrient cycling is its bread and butter. A soil where there is no nutrient cycling, when it is bare or when its nutrients have been depleted, is a dying soil and is vulnerable to erosion and degradation. Conversely, good nutrient cycling implies a biologically active soil operating at peak efficiency and in full health.

MARKETING

It is difficult to emphasise enough the importance of marketing for the organic farmer. This is even more true for the small-scale organic farmer, whose impact on the market and whose profit margins are both likely to be small too. The advantages of organic produce over their conventional counterparts have already been discussed, so what possibilities are available to capitalise on those advantages in our marketing?

At the risk of beginning in the negative, I am first going to register a complaint. Many organic growers have the mistaken belief that the organicness of what they grow should be enough in itself to sell their produce. As a result, their grading and quality control is negligible and they push on to the customer a product that leaves a lot to be desired. I say 'mistaken belief' because you may sell those last few slug-ridden lettuces this year, but your reputation, and worse, the reputation of the organic movement as a whole, will be damaged for next year.

Those few committed organic eaters may put up with the extra protein, but it will certainly put off those potential converts on whose goodwill and support we all depend for our continued success and growth.

In the long run the organic movement needs to reach and attract the average shopper and not be restricted to a small élite who will put up with anything, whether because of need or commitment. This would be so if only because of our philosophy, but hard economic reality demands it too.

We have a quality product to sell. Let that be reflected as far as possible in its external quality as well. Organic farming is no longer in the dark ages of technology or experience, so it is perfectly possible to produce most crops without pest damage or blemishes. If you cannot, then perhaps you should be growing something else.

The next logical extension to this point is presentation. Here we have much more of a dilemma. The cost of good-quality packaging, or even the not so good, is high, and its generally non-renewable nature does rather go against organic principles. Good packaging is, however, essential for both presentation and preservation of your produce, especially if it has to travel any distance.

Recycled wooden boxes can work very well if you have a regular source of them and if they are lined and over-wrapped with paper or plastic. Recycled cardboard is usually disastrous. It seldom stacks, it has lost its original rigidity and it looks awful, perhaps not as it leaves the farm but certainly after its journey to the shops. Its contents suffer accordingly. Reusable cartons and crates are the last option. They are marvellous but expensive, so you have to make sure that you can get them back.

On the livestock side, the situation is more critical. Livestock products are less likely to be selling predominantly on their organic merits, and will be in more direct competition with conventional equivalents. Luckily their very nature requires good packaging and presentation and there is little room for manoeuvre which would lower standards. Quality is not so much a matter of careful grading, but more a matter of cleanliness and good management in the processing stage.

The next important aspect of good marketing is attention to

detail. This really runs through all areas of organic farming, not to say all farming, even all businesses. To grow quality and to carry that through to the customer requires attention to every detail. One bad batch of produce damages your reputation for an awful lot longer than that produce lasts. When dealing with supermarkets, one bad box in a load, even *slightly* underweight or oversized, might very well mean the rejection of that whole load.

What are the options for marketing? These are basically two; either you sell direct to the public or to some kind of middle man or wholesaler. Within each there are many subdivisions.

Direct Sales

You can reach the public direct by several methods, ranging from a farm shop on site, through a market stall or retail round, or even stretch to your own shop in town. You will need a reasonable range of produce to attract customers and, very important, you will need good continuity of supply. We always found that our regular clientele in our farm shop were in direct proportion to the range and continuity of our produce. You may therefore have to buy in other goods from other organic producers, or do a line in conventional produce too.

Any premises selling food, or anything for that matter, is subject to planning controls, which you must find out about in your local area. The only exception is a farm shop selling food that has been grown on the holding, but not processed. There are considerable grey areas within the planners' definition of processing, and they try and draw into the net such things as frozen milk or vegetables, yoghurt, cream and meat, even meat which has been butchered on the farm.

If the planning people do not mind you, the health and safety people might. They are particularly strict, and rightly so, with processed foods, meat and dairy products, and with processing operations too. If you are thinking of starting up anything along these lines, it is always best to get their approval from the beginning. Seek their advice at the outset and you will find them friendly and helpful, and your progress is likely to be much smoother as a result.

The main advantage of selling direct is that your mark up can be higher, with prices pitched nearer the retail level. You are also in contact with your customers, which can be valuable both in terms of feedback and your social life, as farming is a rather lonely occupation. Another benefit is that your packaging can perhaps be less sophisticated, or even non-existent if you go in for the pick-your-own business. You also have a ready market for your outgrades which can be sold off cheap as bargains.

There are however just as many disadvantages. All direct selling requires time, which will leave less time for your actual farming. If you are really serious about selling, it may warrant extra help. Alternatively, if you are not careful, you may find yourself almost entirely taken up with marketing. At times it may be a positive bind, particularly at weekends or in busy seasons.

There are other expenses too. You may well have to embark on a heavy advertising programme in the local press. The cost of converting a farm building and equipping a shop, renting a market stall or paying running costs for a mobile shop, the cost of carrying stock and covering wastage must all be taken into account. You may also have to travel some distances to obtain the right products to go alongside your own on the shelves, and the mark-up that you can put on these may hardly justify the expense of the extra time and work.

These are all considerations involving economics and time, and must be weighed up carefully before you embark on anything at all. The implications for your cropping regime are equally important. You will want a steady flow of as wide a range of produce as you can manage. You will therefore lose out on economies of scale, and will have to concentrate on many different varieties, preferably grown in small lots to get good succession. This is time-consuming, but may suit the smaller, more intensive situation better, where double and treble cropping can be practised.

The livestock products are slightly less demanding in this respect. They are not so seasonal and continuity is easier to attain. The production of free-range eggs or goats' milk on a sustained and regular basis is nevertheless quite an art, and its difficulty should not be underestimated. With sheep's milk, regular production is almost impossible.

Always keep one eye on the competition. There is a plethora

of mobile shops these days, and an increasing number of market stalls sell organic produce or at least wholefoods. Many areas are fairly saturated with farm shops and pick-your-own places, and unless you are in a particularly good position, right next to a town or on a main road, your organic advantage may not be enough of a draw. Unfortunately it still even puts off some people, who would rather keep to conventional produce, feeling safer sticking to what they know about.

Wholesaling

The next step up is to be your own wholesaler, taking your wares round and selling them in larger amounts to shops, restaurants, hotels and consumer groups. The time involved should be less, assuming the distances to travel and the amounts sold are not out of proportion, but you will still need a vehicle.

Shops taking fruit and vegetables will probably want a wide range and continuity of supply, especially if they are trying to develop an organic section. The same conditions apply to your own shop, and very soon you may find yourself running about the countryside collecting other people's produce to keep your shop customers happy. This is less likely to happen now than a few years ago, as many independent local organic wholesalers have set up business and can do the job properly on a full-time basis. Competing with them would probably not be profitable or worthwhile.

One type of shop to beware of is, unfortunately, the wholefood shop. Wholefood shops always want organic produce the most, but are able to sell it the least. This is partly because they are not versed in dealing with fresh produce, and partly because the range that they end up with is not sufficient to attract the customers. However much they want it, they never seem to know how to set up a good display, nor how to keep the produce looking fresh and topped. This is a continuous job and may even involve throwing some away. When the produce does not look good, it does not sell. The wholefood shops are then reluctant to buy more and their range diminishes accordingly.

Greengrocers can be excellent outlets if they offer whole-hearted support. However, greengrocery is a dying trade, with the inexorable advance of supermarkets, and the calibre of many greengrocers explains why. A new venture, like having an organic section, needs attention and enthusiasm. Those that give this are well rewarded, but many do not. On the other hand, many like to have local grown produce and can sell prodigious quantities under that name, which some seem to think is almost synonymous with organic. You may then be heading more towards scale rather than variety.

Further still along that path are the supermarkets, but they are to be regarded with great caution. Their potential is absolutely enormous. They account for about half of all fresh fruit and vegetable sales in Britain and they are all falling over themselves to get on the organic bandwagon. This is so much in contrast to the greengrocers who are so busy missing the opportunity. Most supermarkets require their produce to be pre-packed and the quantities that they need will almost certainly be more than one individual producer can manage. They are much better dealt with through a third party, who can hopefully be a buffer between you when they start rejecting loads, and who can give them the continuity that they demand.

The better hotels and restaurants are usually after quality and unusual or early things. They can be ideal for the organic producer and they are beginning to realise what organics can offer them. This is a potential growth area that has yet to be developed fully. Remember, however, that quality and continuity, if only of a few lines, are vital. This probably means speciality, which could mean herbs, guinea fowl, fancy cheese or whatever they want.

Livestock products may have a greater range of outlets that might be suitable, including delicatessens, dairy shops and butchers as well as wholefood and health-food shops. Their presentation and the way that shopkeepers deal with them are likely to be less of a problem than with fruit and vegetables, but their more direct competition with conventional produce might mean less of an organic premium.

Lastly there are the consumer groups. This is something peculiar to the organic movement and is very much in keeping with its philosophy. Basically, a number of committed individuals come together to pool their buying power so that the grower can deal in wholesale quantities, and the group get the benefit of fresh organic produce at less than retail prices. It very much depends for its success on one active person in the group who can compile a regular order from everyone's individual needs, notify the grower, receive the order, divide it up and oversee its collection or delivery. It demands a lot of time and therefore commitment on the part of the co-ordinator, and involves a fairly close relationship between grower and group so that the demand and supply are well matched (variety and continuity again), but it can be rewarding and beneficial all round.

In Europe and in the biodynamic movement the system is so highly developed that some farms are entirely run for their consumer group. In some cases the group virtually owns the farm and they all sit down together to work out the costs of producing what they want, not only vegetables but also milk, cheese, bread, meat and so on, assuming average yields. They then pay a set amount which includes a salary for the farmer, so that his income is assured even in times of crop failure. It is an idyllic situation, with true participation of the consumer in his or her food production, and no excess pressure on the land to force higher income out of it.

On that idealistic note we come now to the middle men. We may curse them but they are often indispensable. Specialisation works both ways and the wholesaler can spend more time providing a better range and service to his customers than a hard-pressed farmer trying to do too much. At this time I would advise not depending on the conventional wholesalers, except perhaps for your surplus (which they will not like), unless you have a good relationship with one already. Their prices and their understanding of organics are generally low. There are exceptions in some areas with some wholesale markets, but usually it is the quality of produce that makes them so.

The organic wholesalers, on the other hand, are a new and enthusiastic breed. A brand new network has rapidly spread all over the country, in two main tiers: the big importers and distributors mostly in the large conurbations, of which the Organic Farm Foods group is the foremost, and the smaller independent wholesalers serving much more localised areas. They have all encountered their fair share of hiccups and teething problems, but their contribution to the growth and expanding potential of organic farming has been tremendous. The enormous range of imported organic goods has helped to fuel demand further, and the service, range and demand of these wholesalers increases every year.

They can also shift sizeable quantities for you if, once again, you do your homework and co-ordinate your crops with their requirements. They are the main suppliers of supermarkets, and it is a salutary thought that the vast bulk of this has been imported, because there just is not the quantity, or quality of produce in Britain to satisfy even their initial needs.

One thing to remember with fruit and vegetables is that they are all subject to Common Market regulations regarding grading and labelling, except farm gate sales. Some crops grown throughout the EEC, like apples, tomatoes, courgettes, etc., have very stringent grading requirements that involve not only blemishes and damage, but also size and ridiculous things like stalk length. Others, such as swedes and parsnips have no grading requirements.

Organic crops are not exempt and the horticultural inspectors are beginning to discover this rich new hunting ground. Your good presentation must therefore comply with these regulations and must also include correct labelling. Organic produce, of course, needs additional labelling too, to proclaim that it is organic and verifiably so.

Wholesaling the livestock products is a different matter altogether. As yet, there is no organic meat company to fulfil that much-needed function, although one is being planned at the moment. Dairy products are somewhere in between. Some organic vegetable wholesalers trade in them, but the market is still largely undeveloped. You are more likely to have to rely on your own resources.

The Marketing Co-operative

The last and perhaps most important development in marketing for organic producers is the co-operative. They are commonplace in conventional agriculture and are enthusiastically supported by Food From Britain, who are able to provide considerable funds towards the costs of starting up. In fact, FFB are very keen on the potential for organic marketing co-operatives; they see the challenge of a clean slate in the organic marketing world, providing a golden opportunity for co-operative development unfettered by previous structures. Several are now successfully operating.

It is important to understand what is meant by a marketing co-operative. The first point to stress is that it has nothing to do with common ownership or other aspects of ordinary co-operatives. It is only to do with marketing. Members of a co-operative sell their produce together, thus enjoying economies of scale, and co-operation not competition. It gives the producer the security of the best possible market, assuming the co-operative is doing its job, without the hassles and time involved in doing it all himself.

The main advantage of a marketing co-operative for the small producer is therefore that, together with others in the area, he has a combined size that has clout and selling power way above what he would have alone. Very often the prices that the co-operative can get are above those that an individual could manage, even after the co-operative's costs, usually taken on a percentage basis, have been deducted.

Of course, the decision as to what is grown must also be taken co-operatively, for the members should only grow what the co-operative has markets for. This does, however, allow members to concentrate on what they can produce best, thus gaining the benefits of specialisation without the disadvantages of marketing it on a small scale.

Two important points must be remembered. First, the co-operative, although owned by its members, never actually buys or owns the produce itself. It only sells it on behalf of its members. Second, the co-operative should not make a profit. If it does, then its commission charges are too high. Its job is not to make money itself, but to achieve maximum returns for its members.

Finally, one thing above all determines the success of a co-operative. That is the total commitment of its members. Selling the odd box down the road because you can get a few pence more, weakens the overall strength of the co-operative. You would only be in competition with yourself, and you, along with the other members, would all be weakened in the long run. Marketing co-operatives stand or fall on the commitment of their members.

As more and more people get into organic production, the marketing co-operative provides particularly the small grower with perhaps the best and most secure way of marketing his produce. If you are interested in setting one up, there is much help and advice available. It will involve a lot of time and effort, countless meetings and a certain amount of paperwork. Do not be put off by all that. The job is becoming easier as each new one is formed.

5
Protected Cropping

It is now time for some practical details on the production of different crops and livestock in the context of the organic system. A complete guide to the husbandry, production and harvesting of each and every variety in great detail is not possible here, but pertinent information on as many different types as possible will be offered.

The most important details are the special treatment or peculiar requirements that our organic methods demand in the growing of the crop. Further information on conventional cultural practices and more general knowledge about individual crops can easily be obtained from ADAS.

Protected cropping is a speciality in itself. Acres of glasshouses or multispan polytunnels full of tomatoes or cucumbers are not usually suitable candidates for organic husbandry. A few are known and exceptionally good they are too, but the capital outlay involved makes them really out of reach of the small farmer, even though the acreage is tiny. The management of such a unit has to be of the highest quality because the constraints of the system, such as virtual monocropping without the conventional aids, make the balance all the more precarious.

More common amongst organic growers is the practice of having a few polytunnels or a small glasshouse integrated into their system even though it is only a small part of their holding. There are many reasons for this; diversity, increased income from a small area, spreading the work-load and cropping period over a longer time, and variety of produce to sell. They tie in well to the rest of the farm too, with uses such as starting off early outdoor crops, chitting potatoes and even drying and storing onions.

PROTECTIVE STRUCTURES

The less specialised and less intensive use of protective structures will be the main consideration here, which basically means 'cold' crops in tunnels. There are, however, other forms of protected cropping, which are more closely associated with the outdoor crops. An ethical question hangs over many of these systems because they are based on polythene, which, of course, is made from non-renewable fossil fuels. Furthermore, the polythene is not generally intended to be used more than once. There is no restriction on their use, but many organic growers are reluctant to embrace them fully. It is an interesting dilemma which highlights one of the grey areas of organic agriculture.

Plastic Mulch

This is usually black, for weed control, heat absorption, and therefore earliness, and it is most commonly used for strawberries. Another type is black on one side, underneath, and white on top, and is used to delay maturity and so to prolong the cropping season. The soil must be moist before laying these because they are not permeable. The exception is Mypex, which is a woven material and so is also stronger but more expensive.

Black polythene is also popular in tunnels (with 'layflat' irrigation tubing underneath it) as an aid to management of weed control and humidity, and to enhance earliness further. If you start with thicker film such as silage polythene, it can be used again and again over many seasons.

The other common form is the biodegradable, clear surface mulch. The crop is planted through the film, or sown before it is laid in which case it comes up through perforations or slits. It gradually degrades in sunlight over a number of days or weeks, so it does not need removing. It is not very suitable for organic growers because of problems with weed control under it.

Floating Film

The crop grows underneath this, holding it up with its own foliage. It has varying percentages of perforation according to the level of ventilation and temperature that is needed. It is not widely used in organic production because, once again, of the weed problem.

Cloches

Cloches are made of either glass or polythene. Glass ones are now outdated because of their initial cost and the high labour requirement of their management. Polythene cloches are now also becoming outdated in conventional horticulture for the same reasons, particularly as mechanisation and technological developments have moved in other directions. However, they are perhaps better suited to organics than some of their replacements, and they do have a place in some systems.

The main advantage for organic growers is the accessibility to the protected area for weed control. They are also fairly versatile

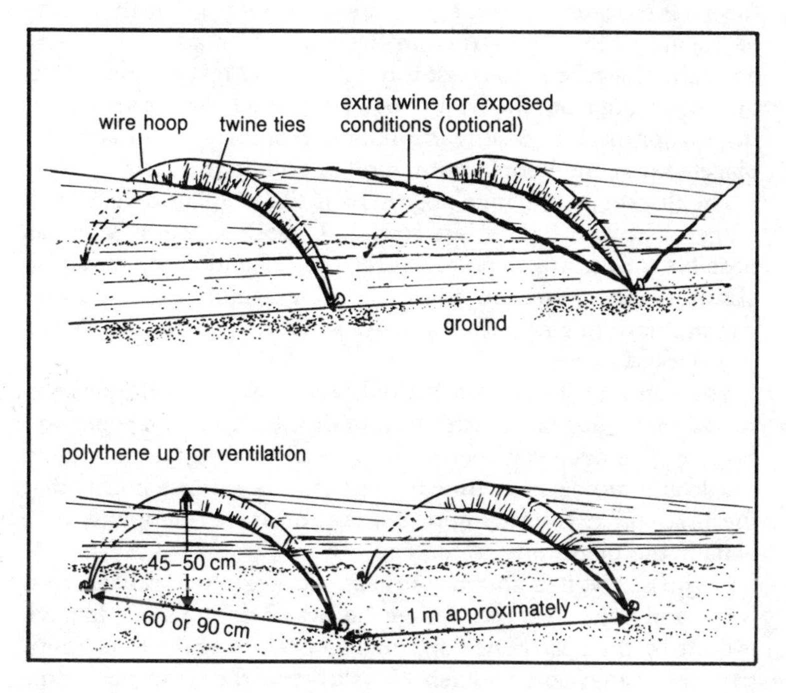

The construction of a polythene cloche.

and the thicker grades of polythene can be reused. The main disadvantage is their susceptibility to wind damage and this must be taken into account when choosing the site. Crops grown under cloches are usually strawberries, lettuce, runner beans, courgettes, tomatoes and also seedbeds.

Polytunnels

Polytunnels provide the cheapest means of starting proper protected cropping from scratch, and probably the best size for general use is the 5.5×18m one, unless your site is particularly exposed (which it preferably should not be), when the 4.3m width might be better. These narrower, and hence lower ones, tend to have more restricted air movement and can get rather too humid. Larger ones still, and multispan structures, are suited to more serious use.

A few tips on construction might not go amiss, as the life of a tunnel is crucially dependent on how it is put up. The first thing to ensure is that the structure itself is well made and strong enough. The tubes must last through several reskinnings without distorting from the weight of a supported crop or succumbing to storms. It is well worth taking great care to get accurate placement of the hoops, with good alignment in the air.

The thicker (800) gauge polythene is likely to last a year or so longer, but has slightly less stretch. Choose as warm a day as possible for putting it on to help it stretch; and as still a day as possible, otherwise you will be battling against the wind and feeling more like an out-of-control sailor or parachutist than a land-locked farmer.

You can now buy foam-backed tape to stick on the outside rim of the hoops on which the skin lies, to reduce its wear still further. The heat that the metal gives off makes the polythene brittle and more susceptible to tearing; it is at these points that the first tear invariably appears, and once that happens, it is usually the beginning of the end.

Surprisingly, it is not the tautness of the skin that causes most wear, and hence tear, but rather the lack of it. The most damage is done by the polythene flapping persistently against the hoop, something that occurs where the skin and the hoop part company near the ground. If it is really taut, this does not happen

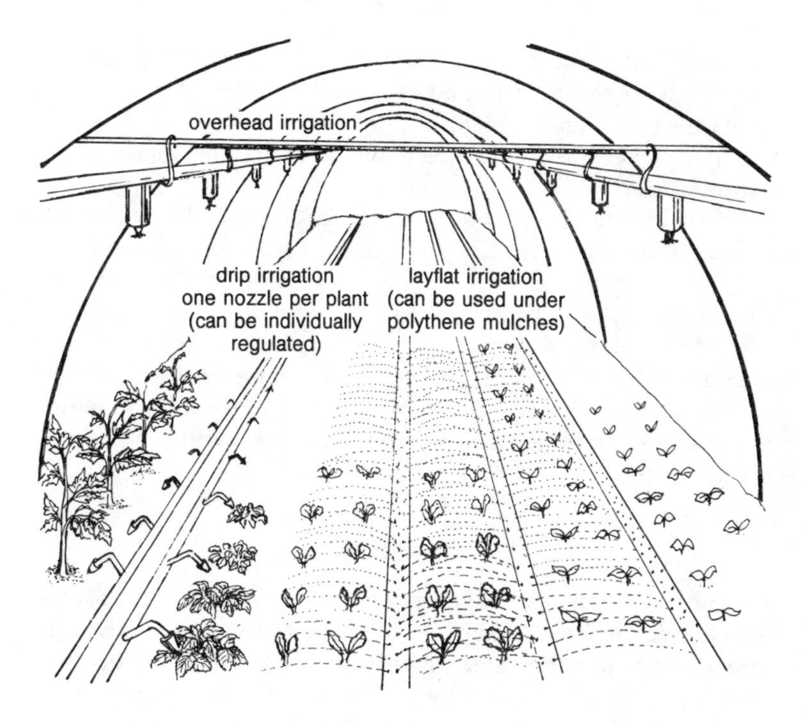

Irrigation in a tunnel.

and its life will be extended by another year or two, until it goes probably further up the hoop. Most designs have a means of jacking up the hoops by about an inch on each side where they sit in the foundation tubes. If you have not got the skin really drum tight, then you can correct it on a subsequent hot day. You are unlikely to be able to overdo this tightening, so you can really put your back into it.

CROPPING

Your choice of crops is likely to be more restricted than out of doors, but the cropping period will be longer. There are three main periods into which the crops can be divided: summer, autumn/winter and spring. The trouble is that most crops will overlap with the next period too much to allow for three full crops a year. Careful planning of the rotation, including the other uses required and green manures too, is therefore an essential start.

Crops in tunnels need not be on a large scale. Their production is intensive and each bed can be regarded separately, as long as the companions in the tunnel are compatible. Four beds fit into a 5.5m width, three into 4.3m, though the two outside ones only take half as much of the tall climbing crops like cucumbers or tomatoes because of the slope of the wall. The main crop is likely to be a summer or spring one, around which the others have to be arranged. We shall start with the summer ones.

Tomatoes

I am not convinced of the value of tomatoes as a cold crop for the small-scale producer. They take a long time to come to anything and when at last they start ripening, the price slumps because everybody else has them too. However, if you need a wide range of produce, they are a must, but they do require quite a lot of care and work. You might be able to get away with the new outdoor bush varieties which seem to crop just as early and almost as much for considerably less bother.

Choose a variety that is well tried and suited to the less intensive conditions, such as Shirley, Arasta, Danny or Ostona. There is not really any well-researched organic seed compost, so a commercial one at this early stage is acceptable. Alternatively a peat/sand mix can be used with about half the concentration of nutrients of potting compost. Prepare it well in advance to allow it to activate. Sow into this six weeks before planting and use heat at 20 degrees Centigrade for germination.

The potting compost is better documented and one presently used by a commercial grower has the following ingredients:

100l	Peat
120g	Humber Organic Manure
20g	Ground Limestone
10g	Hoof and Horn

40g Dried Blood
20g Calcified Seaweed
10ml Liquid Seaweed (concentrate)

Again, prepare well in advance and add a further 20ml seaweed concentrate just before use. This can be used for blocking if sedge peat is used. Variations could include substituting half of the peat with equal quantities of worm compost and composted bark.

Tomatoes are heavy feeders so the soil needs to be equally heavily fed, but their potash requirements exceed that of nitrogen (the ideal ratio is 2:1). To achieve the right balance, a previous leafy crop can be manured, perhaps cucumbers at 200t/ha, or spring lettuce at 100t/ha, or diluted comfrey liquid and seaweed can be fed regularly through the season at 0.25l per plant per week, a good practice in any event. Rock potash should also be part of the manuring programme at a fairly high rate and in good time to be activated. With such high amounts of potash around, the magnesium balance might be upset, so if lime is ever needed use dolomite limestone.

The main disease is blight. As a preventative treatment, 0.5 per cent waterglass can be used, sprayed every week. The last resort is bordeaux mixture, but use it with discretion as copper does not go away once in the soil.

Whitefly is the principle pest, but is well controlled by biological control methods, using the parasitic wasp *Encarsia formosa*. It is important however, to introduce these predators as soon as infestation is noticed as they invariably take longer to build up numbers than the pests themselves.

Cucumbers

Cucumbers make an ideal cold crop as they are extremely prolific. They start cropping much earlier than tomatoes and the price does not generally dip quite so severely. Again, they need careful attention and prompt training or they will very soon grow away from you. Vigorous varieties like Farbiola and Pepinex 69 are best, and the sowing and potting composts can be similar to the tomato mixtures.

Farmyard manure can be used rather than compost, spread at

100t/ha, or double that for a heated crop. Feeding in the mid-season, when cropping really gets under way, is also suggested, and it needs to be richer than with tomatoes. Urine or slurry, diluted to 1 per cent, could be liquid fed along with seaweed. Alternatively other solid manures/fertilisers could be agitated into liquid form or applied solid on to the surface, according to the irrigation system.

The fast, lush growth of cucumbers, coupled with the slightly hotter and more humid conditions that they prefer, tends to encourage mildews and rots. Use a 0.5 per cent solution of sodium bicarbonate against these, possibly with waterglass. Much is to do with careful management and good housekeeping, however. For instance, checks due to irregular watering etc., quickly result in fruits dying back which become sites of infection.

The main pest is red spider mite, which can speedily devas-

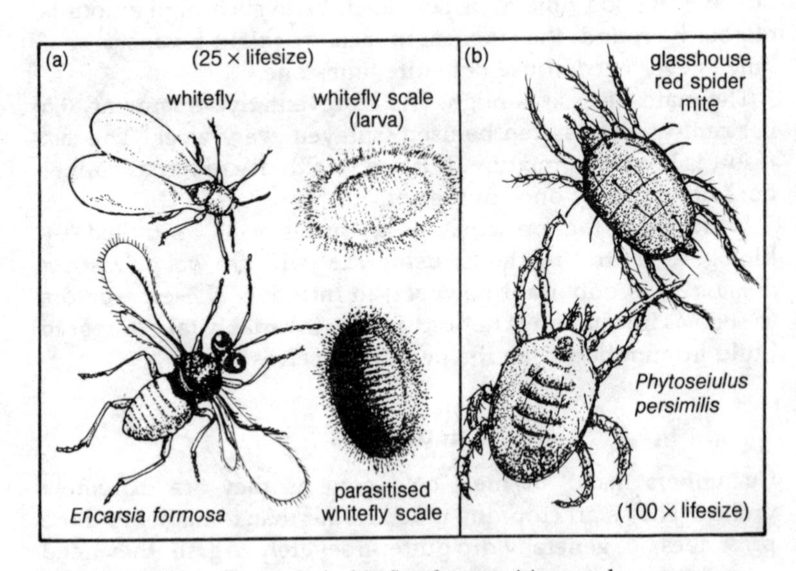

(a) Biological control of whitefly: the parasitic wasp lays an egg in a whitefly scale (the larval form). The egg hatches and the scale turns black. (b) Biological control of the red spider mite: the predatory mite, Phytoseiulus persimilis, *breeds faster than its prey, which it feeds on so voraciously that it can starve itself to extinction.*

tate a crop. Keep good watch and introduce the predatory mite *Phytoseiulus persimilis* at the first sign. Also jack up the humidity a bit. If it is severe, spray with derris or pyrethrum two days before the biological control goes in.

Training methods and spacing of cucumbers and tomatoes is a vast subject in itself, so look at the various alternatives and choose a system that suits you and your conditions. If anything, make your spacings a little wider than conventional ones, as organic soil is unlikely to be as rich a nutrient medium as one fed with chemical fertilisers. The extra space will mean more light and more air circulation, both of which will help the crop and offset any drop in yield that may occur from the lesser density.

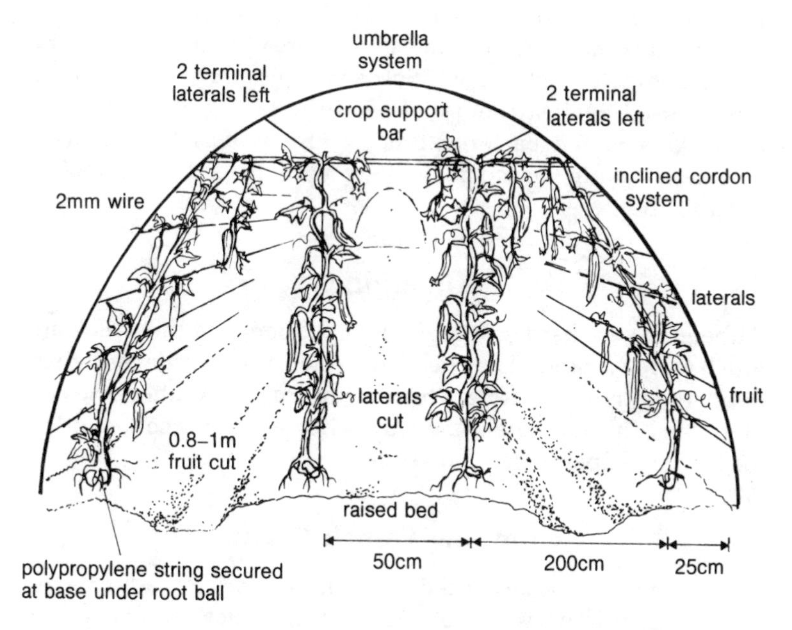

A possible training system for cold crop cucumbers in a 5m tunnel (spacing in row 50cm between plants).

Peppers

Peppers can yield well in polytunnels and, being smaller than the above two crops, are ideally suited to the side beds. Bell Boy is really the only variety worth growing unless you try the more specialist yellow peppers.

The tomato sowing and potting composts can be used, but peppers' growth rate is slower, so they need starting off a bit earlier and they do not need quite so much watering. Manuring of the soil can be similar to tomatoes, but feeding should be more like that for cucumbers as the nitrogen:potash ratio should be nearer 1:1 in the feed. Plant in double rows in the bed with 40cm between plants.

If anything, peppers prefer more humidity than cucumbers, but would make better companions with them than with tomatoes. They also suffer from similar rots and other fungi. Treatment is the same. The most widespread pest is likely to be either aphids or caterpillars. Soft soap, pyrethrum or nicotine can be used against the aphids and the biological control for caterpillars is the bacterial preparation *Bacillus thuringiensis*. Beware of using insecticides along with the other biological controls, however.

Aubergines

Aubergines are harder to grow than peppers and the yields are much more variable. They need to be started off earlier still than peppers as they take longer to come to fruit, but otherwise can be treated in much the same way. They are not recommended for the less experienced, however.

Climbing French Beans

This is a fairly new protected crop, but has several advantages for the organic grower, not the least of which is that it is a leguminous crop and therefore is not so demanding on the soil. However, it is still new enough to meet some resistance in the shops, and prices and demand drop when runner beans start coming in. Its value therefore depends on its earliness.

There are two types, flat podded or round podded. The flat

podded are longer, heavier yielding and easier to pick, but fetch lower prices. Choose one like Selka, Kwintus or Banier. The round podded ones are shorter, like Crystal, or shorter still, like Glastada or Serbo.

The seeds are fairly large and so can be sown direct into larger pots or blocks, two in each, using the tomato potting mix without the dried blood, or an equivalent, less rich compost. Manuring should not be necessary if it was done before the previous crop, as too rich a soil will encourage vegetative, not reproductive growth. However, they do like a reasonable supply of potash which may have to be supplied if it is lacking, although this is unlikely if rock potash is in the manuring programme. Plant in double rows with 30cm between plants in the row.

A second flush of cropping can be achieved in the autumn by stripping off all vegetative growth except the main stem twining up the string, and giving a dressing of fish blood and bone, or other balanced manure, to boost the growing process again.

Aphids and red spider mite are the two main pests, which can be controlled as for the other crops above. Botrytis is the main disease, usually where the atmosphere is too humid. Ventilation is therefore important, but the flowers are fairly sensitive to large changes so care must be taken with it. It is an ideal crop for a black polythene mulch.

Courgettes

In most areas, courgettes are quite good enough outside rather than under cover, but in the less favoured districts they can make a useful protected crop to supply the local market. Here again, earliness is the key factor, particularly as they do not have a pronounced 'flush', so the earlier the cropping, the greater the higher priced early yield.

Like french beans, they can be sown direct into large pots or blocks, but using the full tomato potting compost. Elite is a good early variety. The plants are of the same family as cucumbers, and can receive broadly similar treatment in terms of manuring and feeding. Plant closer together than you would outside, about 60×45cm in two row beds with 100cm between beds. They like it rich and moist, but are less prone to the rots that plague cucumbers, although mildew and sometimes aphids can be a problem.

Lettuce (Autumn/Winter)

Now we come to autumn/winter cropping, which really is dominated by lettuce, with perhaps radish as another possibility. Summer lettuce under cover requires intensive monocropping and is not practised by organic growers, but a useful gap in the market exists with an autumn crop. This can follow the summer main crop, if you do not hang on to it too long, and be ready around Christmas, giving plenty of time for something in the spring.

It is important to remember the growing conditions at this time of the year. They can be summed up succinctly as rapidly diminishing! As November starts to bite, the seemingly reasonable growth of October suddenly disappears and then very little seems to happen until March arrives. The Christmas crop must therefore be almost ready in November. With the low light levels at this time come high nitrate levels in the leaves, because the crop is not able to use the nutrients that it takes up from the soil. It could be argued that we should not be growing leaf crops in the autumn; however, the demand is there, even if the ethics are not.

Sow in seed trays for the smaller scale, or in modules or blocks using a mix as above. The soil should be in good enough heart, if it had summer treatment, not to require any more feeding, but phosphate levels might need particular attention for lettuce. Calcified seaweed at 1t/ha and bonemeal at 0.25t/ha will suffice, providing also other nutrients and slight pH adjustment too.

There are now good crisp varieties for covered cropping, which may be worth investigating, remembering that they take up to two weeks longer to mature. This can be helped by using larger blocks (5cm) and planting them out slightly later. They also need larger spacing, say 25×25cm or more. Marmer, Kellys and Sonia are probably the best at the moment, but the breeders are working hard in this area. Sow in the last half of August to plant in mid-September and harvest in December.

Flats give you a little more leeway with their shorter growing time, allowing early to mid-September sowing and beginning of October planting. Spacing can be down to 20×20cm. Try the dependable Pascal or Ravel.

A good soaking before or after planting should be all the moisture that the crop requires, but a second dose may be needed. Later irrigation tends to encourage the fungal attacks that plague the conventional crop. If problems arise, use liquid seaweed and sodium bicarbonate and watch the humidity. Air heat will help reduce this, and should also be used for frost protection.

Lettuce (Spring)

Much of the above can be applied to the spring crop, but here the growing is initially non-existent and speeds up to great activity. The greatly extended sowing times therefore produce a relatively short cropping period. Continuity can be achieved bearing this in mind, but it does require skill and a bit of luck. Organic soils are slow to get going in the spring, so do not expect quite the length or evenness of succession that the chemical users get.

For flats, start with Ravel and finish with Diamant with the change over at about New Year, using the following approximate guide. Some heat may be required at the propagating stage and for frost protection, but regard it more as another tool for manipulating speed of growth.

Rough Guide for Spring Flat Lettuce Production

SOUTH		NORTH		
Sowing	**Planting**	**Sowing**	**Planting**	**Harvesting**
mid Oct.	late Nov.			late Mar.
mid Nov.	mid Jan.	late Oct.	late Dec.	mid Apr.
early Jan.	mid Feb.	mid Nov.	mid Jan.	late Apr.
late Jan.	early Mar.	mid Jan.	late Feb.	early May
mid Feb.	late Mar.	late Jan.	early Mar.	mid May
early Mar.	early Apr.	mid Feb.	late Mar.	late May

Table 5.1

Crisp varieties can be the same as the autumn crop, sowing in mid-November, January and February, for planting January to early April, and cropping in April and May. If you go for icebergs, Marmer is probably the best, but it will need an extra week or two to mature and should be grown on the slightly wider spacing of 25×30cm. Do not try to boost their growth with more heat as you will probably end up with a looser head.

High temperatures, and particularly extremes between night and day, need to be avoided especially towards harvesting, so use good ventilation at this time. About three weeks before harvest, a second watering, of about 2–3cm, may need to be given. Ventilation will then be doubly necessary, for the conditions will be ideal for the fungal diseases.

The presence of aphids might indicate rather too lush a growth, but can be controlled with derris or pyrethrum. If slugs are a problem, metaldehide mini-pellets can be used, but only in the paths and at the ends. They are likely to be coming in from outside or are hiding in vegetation round the edges. Prevention is therefore better than cure, so they should not be allowed to get into the structure in the first place or reach problem proportions once they are in. Keep the edges clean and keep a barrier of wood ash, bark chips or some such inhibiting material across the entrances. Fertosan have a slug control which can be more liberally used than the others, but it is not so effective.

Strawberries

About the only disadvantage of strawberries in tunnels is that they need to be planted in the autumn for cropping the next spring. You lose a potential autumn crop, but you should not lose any more because they are best not treated as perennials and should be grubbed out to allow a summer crop to follow. It may seem a shame to do away with such good and productive plants in their prime, but economics and efficiency of space really demand it.

Varieties that combine all the elements including taste, so important with strawberries, are really limited to Cambridge Vigour as a good all-rounder and Pantagruella for particular earliness but less yield. The newer Tamella and Hapil are also worth considering, and if you do intend to keep them in over another season then use Redgauntlet for it will make a second crop in autumn.

Strawberries like plenty of organic matter and potash but otherwise are not heavy feeders. Compost at about 50t/ha should be sufficient, depending on past manuring and future plans, worked in to the top 10cm. Water to saturation, put down layflat irrigation tubing and then lay black polythene on top. Plant in September at 32.5cm spacing each way, or 23cm for Pantagruella. Reduce this by 2.5cm per month if you plant later, to compensate for the smaller plants and lower yields that will result.

Ventilate freely, except in very windy conditions until January. Then keep the doors closed, except to clear condensation until flowering, when they need to be open again for the pollinating insects. (Blowflies, bought as maggots from angling shops, can be used if you think there are not enough bees and other pollinating insects around.) Water little and often from April onwards.

Red spider mite and aphids are likely to be the main pests. Biological control can be used for the former, and rhubarb leaf spray or ladybirds can be used for the latter. The need to encourage pollinating insects means that pyrethrum should only be a last resort. Botrytis and mildew are favoured by humid conditions, which means that careful attention to irrigation and ventilation should prevent them from appearing. Liquid seaweed and sodium bicarbonate can be used if needed.

Radishes

The great advantage of radishes is their speed of growth, down to about three weeks in the summer. However it is the sort of crop that requires a good succession. You need to engineer a steady supply, but it will not hold long. There are many varieties to choose from, traditional and new, in both the round and the long, French breakfast types. Try Saxerre, Qum kader or Marabella for spring-sown round ones. With heat and the right varieties, you can be producing radish virtually all the year round, given a suitable rotation. Two crops in winter or three in summer would probably be acceptable as one course.

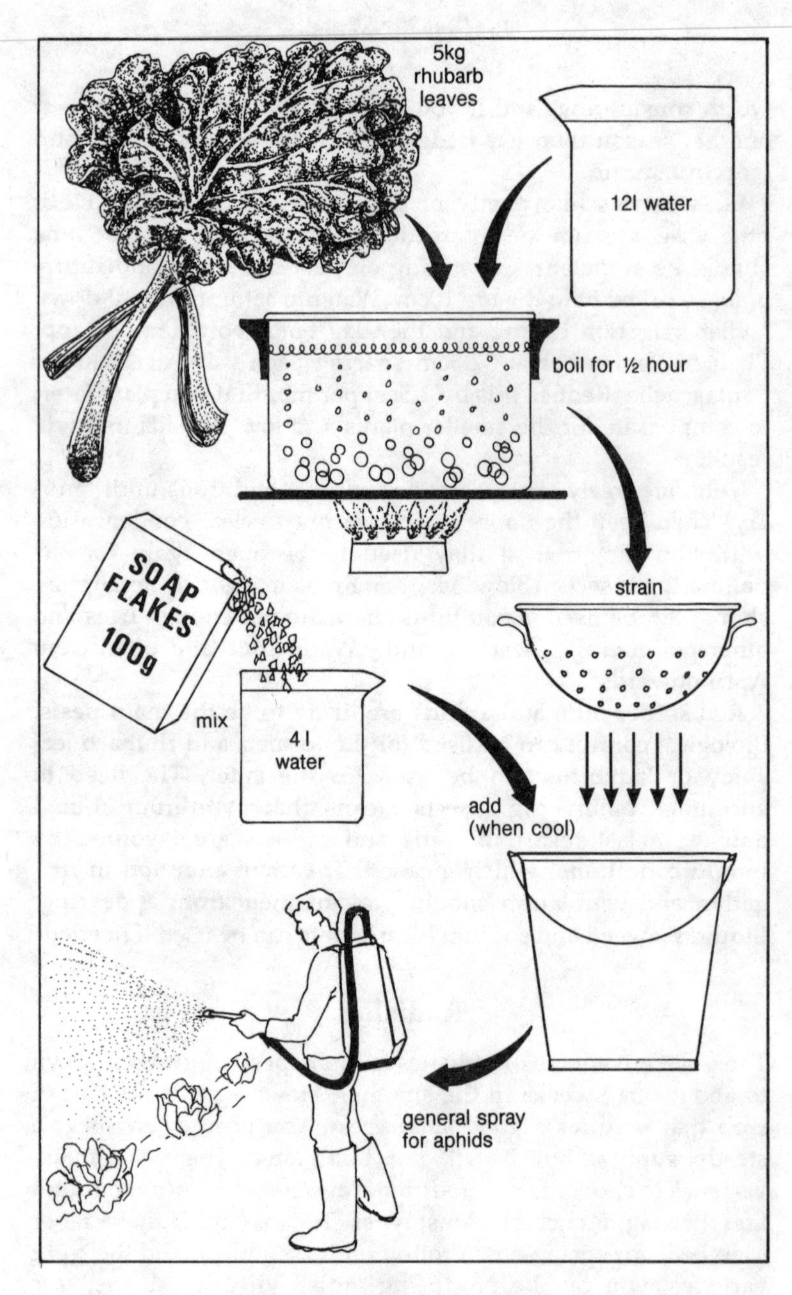

Rhubarb leaf spray production.

Radishes are not heavy feeders, so the soil can probably do without extra manure. Drill them at 2.5cm spacing in rows that can go down to 10cm apart if you have weed-free soil which will not need hoeing. Otherwise go up to 20cm, for if radishes grow fast, weeds grow faster and they will soon smother the crop if not controlled. Water frequently because they are shallow rooted, are susceptible to splitting and they need to be kept moving.

The worst problems you are likely to encounter are mice, which only bait and mousetraps can keep at bay, or a cat, if yours will oblige you, and slugs, which can be dealt with as for lettuce. Root fly and fleabeetle do not seem to affect the protected crop.

Parsley

Parsley is a winter and spring crop. It is slow growing enough to require establishment in early autumn, but it will then crop up to Christmas and again from about February until it bolts in April or May. You have a choice of the moss-curled varieties or the plain, broad-leaved ones. The former are the most common but the latter are widely used in Europe and are both stronger in flavour and heavier in yield and hence are gaining in popularity here. But you may have to do a bit of consumer education if you go for a broad-leaved parsley.

Sow in early July, for pricking out in August and planting in September. In fact you have a month's grace on either side should the rotation demand it, but obviously the earlier it goes in, the more cropping you will get before growth stops. Spacing can be close, down to 15cm or below, and the soil needs to be reasonably rich, although the previous crop's manuring will probably be sufficient.

After establishment, the ground can be on the dry side. This is especially important in the winter months as an added help against the cold, but heat should not be necessary for frost protection.

Spring Onions

This is another winter and spring crop, needing to be sown in September for cropping from February onwards. Choose a winter hardy variety like Winter White Bunching or Winter Over and sow pretty thick in 22cm rows after a good weed strike from a false seedbed.

The soil will probably not need further manuring if it was fed for the previous crop, and watering can be light to medium. The main problem will be weeds which need to be kept well under control, for onions start small and slow. You may get in a pre-emergence burn if you are careful, but the warmth of September may make germination too fast. Pulling can be selective, rather than once over, to prolong the harvest, which in this way will last well into May.

Carrots

The demand for organic carrots is almost insatiable, especially in May and June when there are virtually none around. A tunnel crop is therefore an inviting proposition and can help to fill this gap, either bunched or, later, loose. Traditional varieties of the Amsterdam forcing type or hybrids like Buror or Nandor can be used, and for the best and most even stand, it is preferable to use pelleted seed and a precision drill.

The soil needs to be worked deep and to a fine seedbed. It should not be manured, but it should not need it. Weeds are likely to be the main problem, so the first requirement is a false seedbed prepared at least two weeks in advance of sowing and well watered. Unfortunately not much germinates in November and January, and in this case you want it to, so the longer the false seedbed can be left the better. A bit of heat at this time may seem extravagant but it might pay off if you are expecting a lot of weeds.

Sow at 2.5cm spacings, thicker if not precision drilled, in rows about 22cm apart and make sure that they are accurate, for ease of hoeing later on. You should be able to drill across the whole width, with perhaps the centre row missed out for access, if you are willing to accept a certain amount of damage from trampling.

Water, to settle in, wait until just before emergence and then flame weed. A pane of glass on the seedbed will advance the carrots underneath it sufficiently to give you warning when this will be. Flame weeding should give substantial control but you will almost certainly have to do some hoeing and hand weeding in the rows during the life of the crop, to keep it as clean as it needs to be.

Watering thereafter should be regular and sufficient to keep the soil profile moist at all times, otherwise splitting may occur. However it should be stopped about a week before harvest to allow the pulling of reasonably clean roots. You will then not need to wash, which is quite acceptable to the organic market.

Ventilation is not so important for carrots, and need not be considered until temperatures inside reach 25 degrees Centigrade. Furthermore, always have the lower panels closed to keep out carrot root fly. This is the main pest problem that you are likely to encounter, apart from mice which can be dealt with as above.

Beetroot

We come now to the more marginal protected crops. Marginal because of their less accommodating growing periods, or because of their lower returns. All manner of things have been tried and it is doubtless worth trying many more, for the organic market is not necessarily the same as the conventional. It will accept some crops at prices which might justify special treatment or more expensive methods.

Bunch beetroot takes a little longer to crop than carrots but should return about the same. The stand will be more uneven due to the variation in germination and so will need selective pulling, but fewer roots make up a bunch. Use Boltardy or one of the new monogerm varieties, if they will not bolt and are early. The value of rubbed and graded seed is doubtful.

The soil need not be so deeply cultivated as for carrots, and can be manured lightly in the autumn, say at 10t/ha. Drill in February or March, for harvesting in May and June or June and July, at 5cm spacing with 30cm rows. Pre-emergence weed control, watering and ventilation can all be the same as for carrots. They will also benefit from regular sprays of seaweed.

Other Crops

There are several other crops that can be grown in protected structures. If you want to try something different, here are a few ideas. ADAS, together with your own organic understanding, should come up with suitable methods.

First are other plants for selling, like herbs, tomato plants, bedding plants, and others. There is often a ready sale for these if the quality is good, but do make sure you have your markets sorted out before embarking on anything ambitious. Some of these might tie in with your own plant raising, if for instance you are using modules of some sort.

Indeed, plant raising can itself be a profitable use for a structure and, with good organisation, might be extended to producing plants for other organic growers too. Here you are getting into very specialised areas, even if they are still wide open in organics. Great care and plenty of homework should attend any attempts in this direction. Varieties, quantities (which must include provision for losses), nutrient requirements, germinating temperatures, pests and diseases, adequate equipment, all these things and doubtless more would need your attention.

Next might come the production of cut herbs, such as basil, which is too tender to grow well outdoors. This market is embryonic, even in conventional circles, but is common in Europe and surely set to grow over here. Basil is a greedy feeder and can be susceptible to aphids. Any checks to growth will encourage it to bolt, so it needs good treatment to get good results, but it has the most wonderful aroma.

Another group of crops from Europe that are gaining in popularity here are the different salad crops, and particularly the winter salads. Corn salad, endive, batavian lettuce are all possibilities, and you could even try chicory, grown outside and forced through the winter inside. Other salad crops that might be worth considering include chinese cabbage, a very fast grower that cannot tolerate any checks, and celery, a very slow grower that needs all the help it can get. If you really want to start a production line, I know of no organic producer yet of salad cress.

Spinach is something that is grown commercially with protection, but hardly at all by organic growers. Winter crops are therefore likely to find a good sale. In Europe kohl rabi is widely grown inside, particularly organically. Most kohl rabi sold here is imported, so there is great potential for production of this relatively new crop. However, it goes woody easily and this must be prevented.

Lastly, melons are now becoming a potential crop in this country. Their economics are still fairly marginal and growing systems for them are still being worked out, but it may be worth further investigation considering the high price of imported organic melons.

A method that is worth further thought in the small scale structures is inter-cropping. Plant one quick-growing crop in between where the maincrop will later go, and harvest it before the maincrop grows up to fill the gaps. This is more possible when the pressure on time is lessened by not having crops succeeding each other too quickly, and is a way of making up for a less intensive programme all at once. Lettuce or celery before tomatoes would be ideal.

Doubtless there are plenty of other ideas to toy with. Flowers, for instance are a possibility if an organic market could be developed. Many chemicals are used in their production, and especially those that are banned on food plants, but they still come into our houses and we breathe in deeply their lovely scent.

SEEDBEDS

One good use for tunnels in the spring is as seedbeds for early outside crops. Most of these will be brassicas, but protected cropping rotations do not usually have an excess problem with them. This can therefore make a suitable break from lettuce or carrots. Towards the end of January, spring cabbage and cauliflowers (with a bit of heat) and summer cabbage can be started off for transplanting at the end of March. More cabbage and cauliflowers, red cabbage, calabrese and Brussels sprouts can go in in late February for transplanting in April. Then, in March, early leeks can be sown for transplanting in June. In fact they too could go in in February if you are planning a really early crop.

All of these can be in 22cm rows with fairly close spacing, and in well-fed soil. There are two important things to beware of: firstly, they must not be so thick as to force them into 'legginess', as that will affect their sturdiness and hence their later establishment. Secondly, watch their development closely, for in the heat of a tunnel they will very quickly get too big and that again will affect their establishment.

Water regularly, you do not need to fear fungal problems, and weed well to avoid extra competition causing them to stretch. This is especially important for the leeks, to which you must also try to give a pre-emergence burn. They are slow to come up, are very small and delicate, and in the early stages look too much like annual grass for comfortable weeding. They are very sensitive to competition which in the later stages will cause them to stretch so much that they cannot stay upright. Then you will have a terrible time trying to plant them, all bent and twisted. Remember to water well before lifting and to harden everything off as best you can with all the ventilation available.

6
Intensive Vegetables

The distinction between intensive and field vegetables is, to a certain extent, rather artificial. You cannot entirely divide them by large differences in returns, nor necessarily by labour requirements. Acreages grown may be the same and indeed both types may be found in the same rotation. Having said all that, the distinction is a useful one because in general all these are distinguishing factors, to a greater or lesser extent, depending on the crop concerned. So choice of cropping and growing methods for any particular situation can be more easily determined with the help of this division.

The subdivisions into different types of vegetables are fairly standard – salads, legumes, brassicas and roots. These should be remembered and adhered to when thinking about your rotation. Be aware also of any types that occur in different groups; radish is a brassica, for instance.

SYSTEMS

Intensive vegetables require intensive systems to get the best out of them. Small areas with fertile soil and favoured conditions, on which much labour and attention can be lavished, will provide the greatest returns. It is a general rule that the smaller the acreage the greater the returns per acre.

Bed System

The scale can be sufficient to justify tractor power, and the preferred method is the bed system. Once ploughing or other initial cultivations have been done, the beds are marked out with tractor wheelings, across the line of ploughing, and these are used for all subsequent work. Thus, compaction is restricted to the minimum, but access and efficiency are maximised.

Bed widths ought to be the same throughout the farm for obvious reasons of efficiency, but can vary from 152cm or even smaller right up to 203cm. The most common and probably the best all-round size is 182cm (72in). Row widths within the bed can vary according to the crop, but try to rationalise them as much as possible to the minimum or you will find yourself continually adjusting your machinery to fit each different one. That can be a considerable time waster in the short period when conditions might be just right for, for instance, flame weeding or steerage hoeing. The Europeans often have separate machines set up for each row width or other variation, just so that they can take advantage of this timing factor. They also have very neat horticultural tractors with an additional mid-mounting position. They may look strange, but they are ideal for accurate row work.

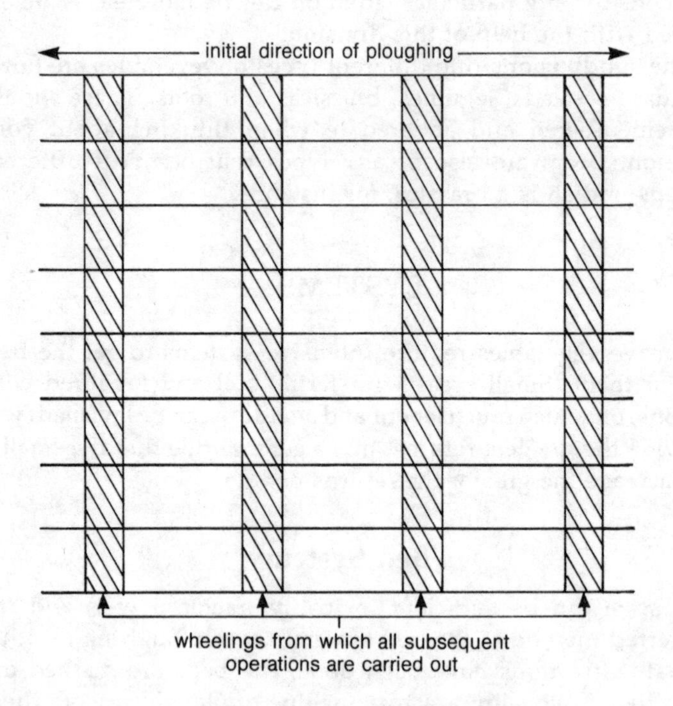

Marking out a bed system.

On a small farm which maybe has only one tractor, the 182cm bed makes for rather a wide wheel spacing for other work. It may therefore be better to come down to a more normal width like 152cm (60in) or even 142cm (56in), and have a semi-bed system based on that. Seedbed preparation, sowing, planting and hoeing can be done as in the bed system, but access for harvesting is limited because there is not enough wheel space between beds once the crop is mature, something which wider beds are designed to accommodate. This is the system that we use, with either one, two or three equally spaced rows per 142cm bed.

Two-wheeled Tractors

The next stage down relies for power on the two-wheeled walk-behind machines of which the Japanese have long been masters. Some of them are incredibly sophisticated, with many attachments, and can do all sorts of jobs. The Japanese have developed them particularly in conjunction with the use of polythene film. The level of their expertise in this puts even the most advanced users in England to shame. For the smaller scale, therefore, these machines are well worth investigating.

Other operations that the two-wheeled tractors are not geared for, will probably require less sophisticated methods. Push drills, hand planting, and more hand hoeing along with inter-row, shallow rotavating for weed control would then be the order of the day. Small, self-propelled block planters and other more mechanised equipment are available and can be appropriate too. It all depends on your degree of intensification, and capital input.

With less mechanisation comes greater versatility. Hand hoeing can cope with any row widths, for instance. So you can tailor your plant densities much more to what is ideal for the crop, rather than to the needs of your equipment, and successional sowings become that much easier. Soil compaction will also be much reduced, so a proper bed system is not really necessary. Remember, however, that with all this comes a greater labour requirement too.

A relatively recent development which could maybe supersede the two-wheeled machines is the new breed of mini tractors. They come with complete systems of their own, all scaled down to a myriad of different sizes. Unfortunately, their individuality and, as yet, their lack of a second-hand market will put them out of the reach of most small-scale organic growers, who would probably do almost as well using and adapting the old Ferguson toolbar system, as many do at the moment. Their forte is in the larger glasshouses, where their expense and versatility can be justified.

Wheel Hoes

A whole system can be built around the push or wheel hoe. This is a very versatile piece of equipment that can be pushed by hand and can cover a far greater area than a normal hand hoe. It can also be used far later in a crop than a tractor and

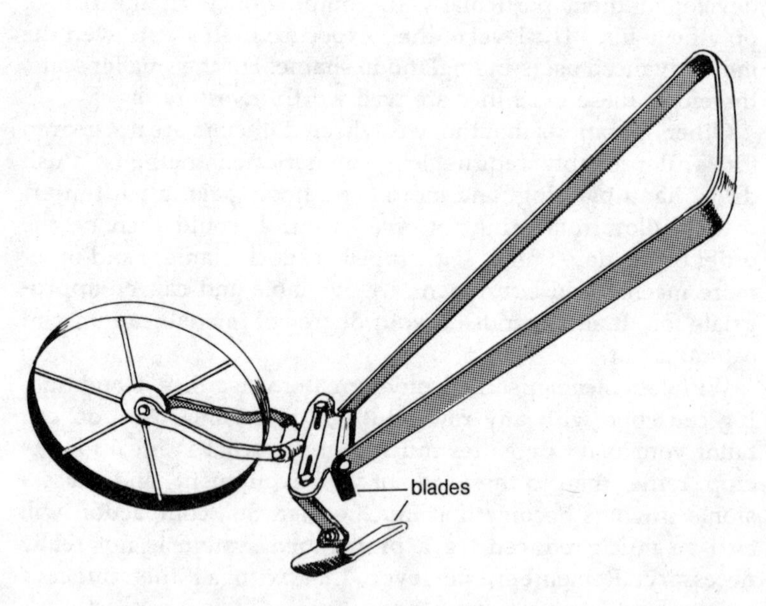

Wheel hoe: the wheel can be moved from side to side to accommodate hoeing on either side of the crop row. Similarly, the blades are adjustable.

steerage hoe. Spacings too can be finely adjusted to the needs of the crop. It is a classic tool of appropriate technology, and is a very useful compromise between mechanisation and hand work for areas that do not justify more than a small Rotavator.

Raised Beds

On the smaller scale still, no machine power is used at all, apart from perhaps for the initial soil preparation. The system uses raised beds and heavy mulching and is the most intensive, and labour-intensive, method of all. Production and returns are correspondingly the highest too, for spacing can be close and double cropping and inter-cropping can be practised.

The beds are formed by double digging (going down two spits) or other deep cultivation, and throwing up the soil from the paths so forming long, low mounds. They should be about 150cm wide to allow easy access by hand without having to step on them. Never being walked on, they should never need digging again. The raised nature of the beds gives them a greater surface area which helps the soil to warm up and to drain. The heavy mulch keeps down weeds and the soil underneath has ideal undisturbed conditions for earthworms and other biological activity, which this system particularly encourages and on which it particularly depends. In all, this is a near perfect environment for optimum growth.

Mulches normally used are either of natural origin, such as old straw, hay or grass clippings, in which case they must be put on a good 15cm deep to ensure that drying out and decomposition do not allow gaps to appear; or they are bought-in materials such as peat or bark chips, but these are expensive; or processed natural materials such as Hortopaper, made from peat fibre; or synthetic such as polythene. Hortopaper is quite new and can be very useful, but it does require careful management. Soil on top of it will rot it faster than anything, it tends to shrink as it dries after rain, and it can easily end up adorning the nearest hedge. It is also quite expensive, but it can be of benefit in some circumstances.

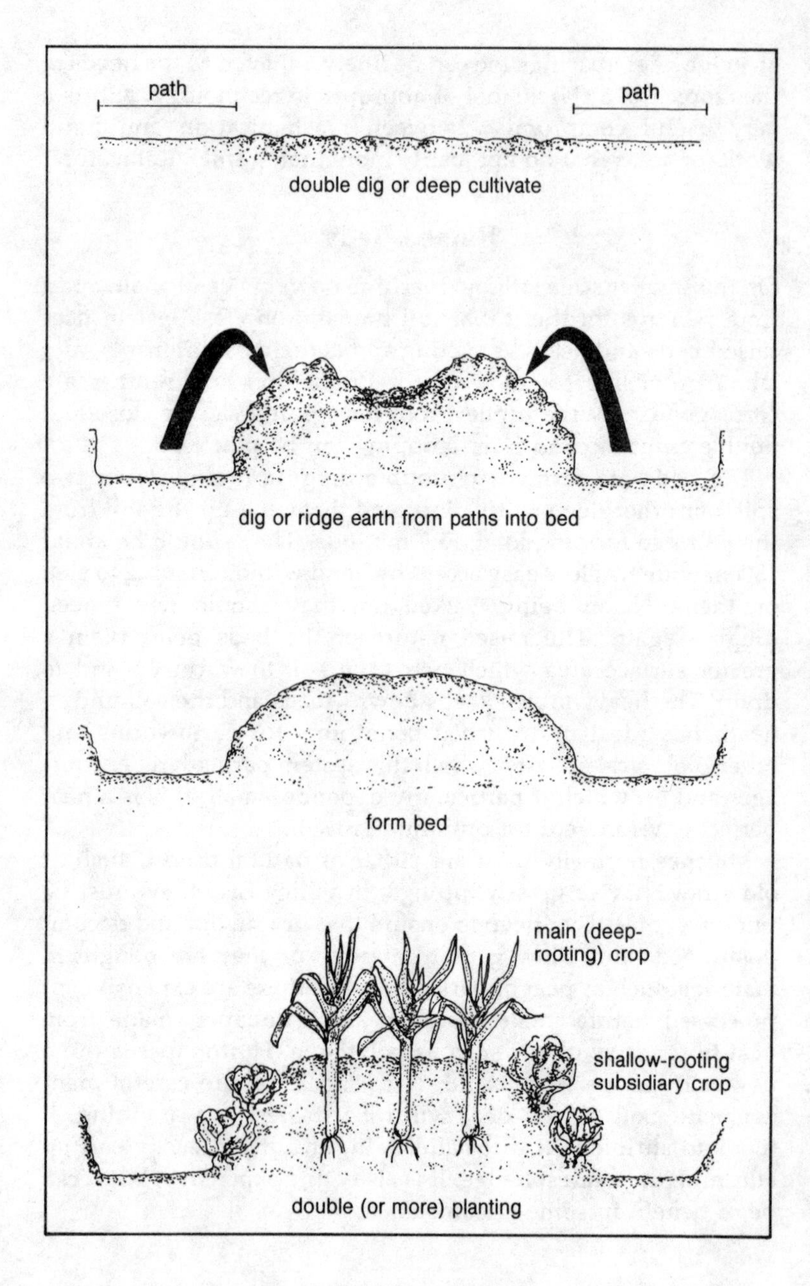

Raised bed cultivation.

The Horse

Gearing up with horse-drawn equipment is another, somewhat tangential possibility. I would not consider it as being a whole system in itself, for it is too limiting, but as an added option in the right circumstances it can be very useful. For instance, horses can go where tractors may not be able to, and in conditions that would otherwise cause damage.

Three things need bearing in mind when dealing with working horses. First, they need to eat a lot, especially when working hard, and they will take up valuable and perhaps limited land for grazing. Second, they demand plenty of care and attention if they are to remain in good condition, even when not being worked. Third, using working horses is an extremely skilled business which should not be underestimated if it is not part of your repertoire.

The horse must nevertheless be one of the most romantic and rewarding sources of power on the farm, part of anyone's dream who aspires to an alternative agriculture, though unfortunately so often shelved when hard reality hits home. There are some fine examples of farms run with working horses. It is possible, and can be profitable, especially if you can turn it into an enterprise in its own right, for example either by breeding heavy horses or using them as a draw to pull in the general public.

SALADS

Lettuce

Lettuce is one of the most important of outdoor crops, and can be divided into three main types: butterhead (flat), crisp (and iceberg) and cos. The last has a subdivision, which is the sweet cos. Little Gem is an example. The fourth type are loose-leaved, like Salad Bowl, and are not grown commercially, although there might be potential market possibilities as different European salad ideas become more popular over here.

Crisp lettuces are beginning to gain the major share, but cos lettuces are also important, more so in the organic than in the conventional market. Both can be treated much the same, as they grow slower than flats, except Little Gem which is small and grows fast. Iceberg lettuces are basically large, crisp lettuces, trimmed to the firm heart and overwrapped. They take about a week longer than the ordinary crisp.

Lettuces are very demanding both of growing conditions and growing skill. They need to be kept moving, particularly flats. In the height of the summer, sowing to harvest can be less than eight weeks, allowing no time for checks to growth. Once ready they need to be cut within a few days. But you also need a good, even succession, providing continuity of supply to your outlets – preferably despite the vagaries of the weather which affects both growth rate and demand. If you can achieve all of this then you will be doing extremely well.

Soil needs to be rich, with reasonable supplies of phosphorus and of neutral pH. Up to 50t/ha of well rotted farmyard manure can be used along with up to 1t/ha calcified seaweed. Flats particularly may also need some quicker releasing nitrogen as well, depending on the state of the soil. Cultivate the beds to a reasonably fine and level tilth.

Blocks or other modules are really the only way to start lettuce off these days, for better timing, establishment, uniformity and weed control. Block raising is a specialist job though, which should not be undertaken lightly unless you can gear up and devote sufficient time to it, in which case it could become another enterprise for the farm. Timing is crucial, but the exactness of conventional plantings is not quite so necessary for the more natural and hence varied organic conditions. Tables 6.1 and 6.2 might serve as very approximate guidelines.

There are a host of varieties to choose from with more coming out all the time. Try Clarion or Capitan for flats and the standard Lobjoits for cos. With crisp, try Jarino early and Nabucco mid and late. Saladin or the other two will also make icebergs. Plant out at 25×30cm for flats and cos, and 30×30 for crisp. Make sure the blocks go out well moistened and their tops stand just proud of the soil to discourage pointed bases and basal rotting.

Lettuce is one crop that really needs irrigation in order to ensure consistent quality. Irrigate to field capacity a few days before planting if necessary and thereafter as required, until the

Continuity Programme for Flat Lettuce

Sow	Plant	Harvest	Cropping period (days)
7 Mar.	10 Apr.	7 June	13
30 Mar.	28 Apr.	16 June	12
22 Apr.	21 May	30 June	10
10 May	7 June	12 July	8
22 May	16 June	23 July	9
6 June	28 June	5 Aug.	10
18 June	10 July	17 Aug.	12
3 July	25 July	3 Sept.	14
20 July	12 Aug.	21 Sept.	15
28 July	20 Aug.	8 Oct.	18

Table 6.1

Continuity Programme for Crisp Lettuce

Sow	Plant	Harvest	Cropping period (days)
19 Jan.	20 Mar.	10 June	20
31 Mar.	1 May	28 June	15
22 Apr.	20 May	14 July	15
18 May	14 June	31 July	16
7 June	29 June	16 Aug.	18
21 June	13 July	2 Sept.	20
3 July	26 July	23 Sept.	24

Table 6.2

leaves begin to meet some ten days before cutting. Then you can stop. Again, flats are the most sensitive ones.

Weeds can be another source of stress. Make sure that they are controlled right from the start, especially chickweed, and keep on top of them. They too will take advantage of the ideal moist conditions that the irrigation produces, so it is likely to be an intense, if short, battle, but nevertheless a must.

The main pests will probably be aphids and slugs which can be controlled as for the indoor crops. The great number of rots that affect lettuces are often brought on by problems with the weather such as damp and muggy conditions. Good housekeeping to clear possible sites of infection, normal fungal control measures such as seaweed, waterglass and sodium bicarbonate sprays, choice of vigorous and resistant varieties and careful management should mitigate the worst effects.

Radish

Radish would make an ideal catch crop, were it not for the fact that it is a brassica which would probably not fit in with the rest of the rotation, that like lettuce its sale needs continuity of supply, and also that it does seem to be prone to the pest problems that the protected crop largely avoids. It is therefore best with a planned succession, rather like lettuce, but from sowing to harvest is more like a month in high summer, with a cropping period of almost as much again.

Crystal Ball or Cherry Belle are good standard varieties that will do all summer long from March onwards, although the first sowing would benefit from some protection. Two other types of radish need mentioning. The first is Japanese radish, rettich or mouli and is not for early production as it tends to bolt. It is gaining in popularity here as an autumn- and winter-stored, cooked vegetable, sown at 15×22cm or equivalent with the last crop lifting at the end of October. Following hard on its heels is the second one, black Spanish radish for winter storage from a July sowing and at similar spacing. Both would obviously need to take their proper place in the brassica rotation.

No special treatment of the soil should be necessary as long as it is in good heart, but they do need to keep moving so irrigation may be needed for the serious producer. Spacings can be virtually anything you can manage depending on how it fits into the rest of your system. It can obviously be a very intensive crop and would be ideal for raised beds and intercropping.

Fleabeetle can decimate the early sowings. A pyrethrum/derris mix is the best remedy, along with a tonic of something like dried blood at 250kg/ha for the plants. If you are troubled by cabbage rootfly there is very little you can do as the roots will have been made unsaleable.

Spring Onions

This slow-growing crop has no place in weedy holdings, but on cleaner ground it is a useful and profitable possibility. Again, successional sowings are needed but not with the intensity of radish or lettuce, perhaps every three or four weeks from March to August, using a winter hardy variety for the last sowings. White Lisbon still seems to be the standard summer variety.

Fertile soil, well manured previously, should be sufficient. The seedbed needs to be fine and firm and prepared sufficiently in advance to allow a weed strike before drilling. The ideal conventional density is about 430 plants/m^2, in rows 30cm apart, but in the organic situation the rows might have to be wider or the density higher, to facilitate and compensate for hoeing. Certainly a pre-emergence burn should be planned.

Another possibility is to sow in multi-blocks and transplant. You are unlikely to achieve the same density unless you hand plant, but it does overcome some of the weed problems. It will also be more expensive although it does mean a much quicker turnaround of crop. Some growers plant a thick onion crop and then use the thinnings as spring onions. It can work on a small scale but is not for the producer of continuity.

Irrigation is largely unnecessary except perhaps as an aid to germination, but weeding is important and needs careful attention, especially in the early stages. Pests and diseases should not worry the organic crop. White rot is soil borne and requires very long rotations if it becomes severe. Leaf spot is more likely in early or winter crops as it is favoured by cool, wet, foggy weather. The standard fungal treatment along with a tonic for the plants should be given.

Parsley

This is another crop only suited to clean holdings, but from an organic point of view it is always best transplanted as its germination is painfully slow. Much of what is in the previous chapter on parsley applies to the outdoor crop, but planting should be planned for March or April. A second planting for overwintering can be done in August, but the protected crop is probably a better bet for this time unless you have space in a reasonably sheltered position outside.

Give the soil a liberal dressing of farmyard manure, say 25t/ha, work it in well and create a reasonably fine tilth, for parsley has a long, if slender, tap root and likes good conditions. Spacing can be down to 7.5×30cm or the equivalent which really means hand planting.

Bush Tomatoes

Bush tomatoes suddenly became fashionable a few years ago, but are now settling down to their more rightful place, which is basically as a good pick-your-own crop in southern areas and of marginal use otherwise. Their yields can be substantial but they are very labour-intensive to pick, and the fruit can be of poor quality, apart from the taste which is usually outstanding.

Alfresco is the highest yielder but not the earliest. Sow at the beginning of April, either into modules with a bit of heat, or pre-germinate the seeds and put straight into (protected) soil. Harden off and transplant about mid-May at $50cm^2$ or the equivalent. For pick-your-own this will have to be modified considerably, more towards a bed system, to allow better access. They need a fertile soil with good supplies of phosphorus and potassium but not too much nitrogen. A moderate dressing of compost and possibly rock phosphate and rock potash should therefore be given and worked in.

Bush tomatoes are an ideal crop for using with plastic film, either over or under it. They start yielding just when the late summer weed flush comes, which causes trouble with fruit quality and with picking. Slugs and blight will probably be the main problems. Clean ground will help the former, but the latter can be more difficult. Ensure first that they are planted well away from any potatoes and preferably upwind. Then do as for protected tomatoes, using Bordeaux mixture as a last resort. In a bad year, blight can decimate what would have been a good crop.

LEGUMES

Broad Beans

Here we have a crop whose returns do not really put it in the intensive bracket, but it is not generally grown on a large scale for it is quite labour-intensive to pick. This makes it a good pick-your-own crop, if one accepts the damage and waste that will result. However its value in an intensive rotation is obvious, both as a legume and as a winter catch crop, and in small amounts it sells well. Aquadulce Claudia is the standard autumn variety and the choice is open for the better flavoured spring sowings.

Broad beans need good supplies of potash and a neutral pH. A small dressing of compost, if you have sufficient, and seaweed in some form will not go amiss. For a proper succession drill 7cm deep into a fairly rough, but well-drained seedbed in October and November (any earlier and the flowers will probably get frosted in the spring) and again in March and April. You can go on sowing until June if you have an assured market to take you through the drop in demand when the runner beans come in. Rows can be anything from 45 to 75cm apart with 10 to 5cm between plants.

Weeds are likely to be a problem in the winter crop unless they can be hoed or ridged to get on top of them before they get ahead in the spring. A false seedbed, for which there should be plenty of time, will help. Weeding in the rows is very laborious due to the close spacing and tillering, but picking a weedy crop is also laborious – you need to decide which is worse. The spring crop should not be a problem. There may not be time for a full false seedbed, but the crop can be chain harrowed before emergence and kept clean thereafter.

Winter-sown beans are susceptible to the chocolate spot fungus, but damage is usually limited and late in organic crops, and should not be a problem. Spring-sown beans tend to get blackfly, but perhaps less so on a field scale. Usually the best advice is to do nothing, except perhaps pinch the tops out if your scale is small enough. A week or two later the natural predators, of which there are plenty, will have the situation well under control. The permitted insecticides can be used if you are desperate, but if given the chance, a blackfly infestation will result in a classic example of natural biological control at work.

Runner Beans

Runner beans are a true intensive crop for they are very labour-intensive but also give very high returns. They are grown to get an early yield from pinched plants using cloches or film, or up sticks or strings for the greater maincrop. They are also popular as a pick-your-own crop. Bokki is probably the best variety as it is stringless, early and high yielding, but Achievement, Enorma and Streamline are also good, the last two yielding over a longer, if later, period.

Runners like good moisture-retaining soil which, depending on type can therefore be given up to 75t/ha of rotted farmyard manure in the autumn or early spring. Potash levels also need watching. The soil needs deep rather than fine cultivation, and the position needs to be sheltered, for wind can damage both the support system and the leaves and pods. It is also advantageous to have bees nearby for pollination.

Runners for pinching need to be early. The quality is not so good, so they have to be well ahead of the maincrop. In fact they are often regarded as a catch crop, being stopped in early August in time for something else to follow. They are therefore best transplanted using modules which can be sown as early as March. Spacing should be about 15cm between plants and one row per 60cm cloche. Surface films, even with perforations, cannot really be used without irrigation under them. Pinch out the growing points when the plants start to run and continue regularly every nine days or so. The more this is done the earlier and better quality the beans will be.

The maincrop cannot be sown until the soil rises above 10 degrees Centigrade, about mid-May. Alternatively they can be started off like the early crop, but with two rows in wider cloches. The film is progressively slit along the top for ventilation and to allow the plants to emerge up the poles. Spacing has to be geared to the support system. About the simplest is two rows of bamboos, inclined together and tied by twine that goes the length of the row and is fastened to a post in the ground at each end. The rows should be about 60cm apart, more in the pathways between each pair, with poles 30cm apart and two plants per pole. You will have to help the running shoots find their supports, remembering that they go clockwise up the poles.

Runner beans are a thirsty crop, very much helped by irrigation. The two critical times are just as the first flowers are opening and again as the pods are swelling, but in dry years and on light land more regular irrigation may be beneficial or even necessary. Once the beans are ready you must pick at least three times a week and very thoroughly too, for old beans go stringy quickly and will take from the developing crop.

Pests and diseases should not worry organic runner beans. It is generally a fairly healthy crop, but you may have to watch out for rabbits who are partial to the young shoots.

French Beans

French beans are, if anything, a little more delicate than runners. Sown monthly from April (with protection or in modules) until early July, they will make a fine crop that will continue through the summer. The later sowings can fit usefully after some early crops have finished. Yields can be high and returns good but they must be harvested very regularly. Furthermore, the beans are smaller and lighter than runners and are therefore even more labour-intensive to pick. The round-podded sorts are more popular and each plant breeder has his own varieties, so the choice is open.

They like a light, warm soil, but it must still be reasonably rich. Apply compost (preferably) at about 30t/ha unless the soil was well fed for a previous crop, checking also for pH (6.5) and potash levels. Make a slightly finer seedbed than for runners, and for the earlier sowings let it warm up as much as it can. Cloches or plastic mulches can be used with advantage in this context as long as moisture content is taken into account. Sowing can be similar to runner beans with 15cm between plants and 75cm between rows, or something equivalent.

Irrigation is as important as with runner beans, and the flowers, being self-pollinating, benefit from being sprayed if the conditions are dry and less conducive to setting. Weeds in such a ground crop are a likely cause for concern, so hoe regularly, or make more use of black plastic mulches.

Pests and diseases are much the same as for runners and should not really be significant. Perhaps the greatest problem is likely to be the quality of the pods, with rain splash, slug damage and stringiness from excessive maturity being the three main contenders.

BRASSICAE

Summer and Autumn Cabbage

I include the summer and autumn cabbages, but not the others, because they really need to be grown successionally, whereas the winter and spring ones are divided more by type, are usually planted all together and are grown more on a field scale. There is overlapping at both ends however, as the early-summer cabbage are pointed and some of the autumn ones are dutch white type.

All cabbages are a real staple both of the British diet and as a crop for growers. The yield is high, the returns can be substantial and they are relatively easy to grow. There are four general groups containing any number of varieties, some appearing in more than one group. A sample of the better ones appears in table 6.3. The first two groups are the early-summer pointed and the early-summer round cabbage. Both can be sown, with heat, in late February for transplanting in the second half of April. Group three are the summer cabbage for sowing at the end of March and transplanting at the end of May. Lastly come the autumn cabbages, sown early May and transplanted in June.

Red cabbage can be treated as being either in groups three or four, depending on variety. Autoro and Ruby Ball are both good.

There obviously needs to be more flexibility in sowing times than is suggested above in order to achieve continuity. Two or even three sowings per group may be required. Alternatively, choice of varieties with different maturity times may arrive at the same result. It rather depends on how you wish the workload to fit into your system. However, do not forget that the

Summer and Autumn Cabbage Varieties

Group	Variety	Planting–Maturity	Average Weight	Standing Ability	Density
1	Hornspi	8½ weeks	0.46kg	poor	loose
	Cape Horn	10	0.69	fair	dense
2	Golden Cross	9	0.54	fair	quite dense
	Balbro	12	0.91	quite good	quite dense
3	Green Express	9½	0.89	quite good	very dense
	Stonehead	11	0.76	good	dense
	Minicole	12½	0.81	very good	very dense
4	Balbro	10	1.26	good	very dense
	Pedrillo	12	1.04	extremely good	very dense
	Erdeno	15	1.72	extremely good	very dense

Table 6.3

early cabbages especially can go over and split, or just get too large and become unsaleable very quickly, so do keep on top of the crop.

Cabbages are heavy feeders, particularly of potash, but too rich a food supply may make them less hardy and resistant. A moderate 25t/ha dressing of manure on reasonably fertile soil should be ample, together with 500kg/ha of rock potash if potassium is on the low side. They also like lime, or rather the clubroot fungus does not, so this is a good time in the rotation to put on lime if it is indicated. In fact it is even better to lime the previous crop, as the lime will then be at maximum effect for the brassicas.

A fairly fine but firm seedbed is required with allowance for a weed strike before planting. Thereafter the crop will probably need one hoeing in the rows and a few between them. If done mechanically at a good speed this can have a slight ridging effect which will help smother the weeds in the row. Irrigation should not be necessary, especially with modules, but can be a boon if you are having to establish bare rooted transplants in hot, dry weather. Thereafter they are best left to fend for themselves.

Flea beetle is the first pest that you are likely to encounter, for which derris and pyrethrum mixed is the best answer. They are tough little things and can decimate the young crop in some seasons. There is very little you can do about cabbage root fly, except accept the few losses that might occur. Bare rooted plants are more susceptible and irrigation can help them to overcome the damage and grow on out of it. Caterpillars are no problem with the new biological control *Bacillus thuringiensis*, but remember that it works only by contact action.

Bunch Turnips

Turnips come in two phases, the summer bunching ones and the maincrop. The former are grown intensively, as successional sowings throughout the summer are required for continuity. The latter are more of a field crop (*see* Chapter 7). The all-white varieties are best for summer work, and the standard one is still Milan White Top. A new hybrid is also available called Tokyo Cross.

The soil needs to be firm and fertile, manured the year before for a previous crop, with a pH of about 6.5. The only other consideration is boron, which may need adjusting with borax at 10kg/ha, for turnips are very sensitive to its deficiency. Sowing can start as early as the end of January inside, moving outside in April and continuing about every three weeks until early July. The bed system is ideal and spacing ought to be about 5×30cm to make small roots that go about six to a bunch.

Turnips are fast growing and need to be kept moving, otherwise the roots lose their crisp tenderness. Irrigation is therefore important if conditions are dry and hoeing must keep well ahead of the weeds to ensure that there is no competition. Flea beetle and cabbage root fly are likely to be the most important pests. Treatment for the former is as above, and needs to be prompt on the young seedlings. Root fly is a real problem and may cause you to abandon bunching turnips if your area is bad. The protected crop will probably escape damage.

Kohl Rabi

This strange, sputnik-like plant is becoming increasingly popular, having once just been used as a stock feed replacement for swedes in the drier eastern counties. The main disadvantage is its tendency to go woody, associated with poor conditions leading to bolting, and a temptation to let it grow too large. Tennis-ball size is quite sufficient, then it should remain succulent and juicy. Try a new variety like Lanro. The purple ones are more for the winter and are not such good quality.

Kohl rabi like light soil but with plenty of body. Farmyard manure at 25t/ha should suffice, and, as this is a brassica, do not forget the lime status. Sow into a reasonable seedbed, after a weed strike, every three or four weeks from March to July, for once again, continuity of supply is important. Spacing needs to be about 35×20cm.

This crop grows quite fast, but needs to, in order to be tender, so hoe well to prevent weed competition. Irrigation should be unnecessary. Pests and diseases, if any, will be the same as the cabbages', but on the whole kohl rabi is a remarkably healthy plant.

ROOTS

Bunch Carrots

Little needs to be added to what was said in Chapter 5. The conditions outside are obviously more variable than they are with protection, although this can be somewhat offset by the use of floating films, which suit carrot production well. Going only for the early bunching market, it makes a useful crop that fills a big gap through the summer after the tunnel crop has finished and before the maincrop comes in. Nevertheless, a succession is important, which can be achieved by sowing times and the use of mulches.

Manuring, soil preparation and plant density can be the same as for the protected crop, although the rows may need to be a little wider apart to facilitate the greater amount of weeding that will probably be needed outside. It is very important to take a good weed strike before drilling and to flame weed pre-emergence. The first sowing in March should only be with some protection. Sow again in April and May. If carrot root fly is prevalent in your area, it is best not to tempt fate and steer away from early carrots.

Bunch Beetroot

Again, bunch beetroot has largely been covered in the last chapter. However, one big difference must be mentioned. The early outdoor crop works well when planted as multi-blocks, and can be made earlier still by covering the blocks with floating film. Sow the beetroot, preferably a monogerm variety, five or six to a peat block with heat at the end of February and plant out at the beginning of April, or earlier if conditions allow, at about 35×20cm spacing. A second and third planting, timed to go out in May and June, should give a steady crop throughout the summer with selective pulling.

Fennel

Here we have another example of an exotic crop that has become more widespread in recent years. The demand for it is still patchy, unless you are hooked into the quality restaurant trade or something similar. Furthermore, it will not keep, either in the field or in store, as it tends to bolt at the slightest opportunity. Fennel requires good, unchecked growing conditions with both sun and moisture if it is to produce a quality crop. Zefa Fino and Zefa Tardo are the standard, but not the only varieties, for early and maincrop production.

The soil is better on the light side as long as it is fairly rich and holds moisture well. Thus compost would be better than manure, spread at 15t/ha. The preferred pH is below 7. The seedbed needs to be reasonably fine and warm. Sow at the end of April and monthly thereafter, changing varieties for July and August. Another possibility is to use pots or blocks, but the seedlings and roots are quite tender and must not be damaged in the process. Ideal spacing is about 25×50cm.

Weed control is very important at all stages of growth, so hoeing and inter-row work must be kept up. Irrigation is also very important if conditions are dry, or the stems will not bulb up. However, pests and diseases are minimal and should not bother the organic grower.

Celery

The potential for organic celery is wide open. Almost none is grown here, with imports coming particularly from Israel. Perhaps the reason for this is that it requires skill, care and a rich soil to grow well. Leaving aside the white types that need earthing up and deep, peat soils, the standard self-blanching celery is Lathom Self-blanching.

The soil needs to be rich and deeply worked with farmyard manure incorporated at up to 75t/ha. PH can be on the acid side at 6 or just below. Sow the seed with heat in March, for pricking into blocks and transplanting at about 25×30cm spacing in early June. You can try for a second sowing, but the frost may well get to it before it is ready. Keep the weeds down and irrigate if the soil gets below the 25mm deficit line. It is slow growing, does not like dry soil, and must be kept moving. A few weeks before harvesting, you can mulch the wheelings between the beds with straw to give the outer rows extra blanche.

The main disease of celery is celery leaf spot. It is seed-borne and can be a big problem. All seed is treated to control it, but this is obviously not allowed in organic farming and so other remedies must be employed. Holding the seed in water at 120 degrees Centigrade for 25 minutes just before sowing should kill the disease without killing the seed. Bordeaux mixture, sprayed on the growing plants at regular intervals should keep the disease at bay if it is established. Good housekeeping thereafter, removing and burning infected leaves, along with the usual tonic measures, should help prevent further spread of the disease.

The celery fly, or rather its larvae, may also be a problem. They are difficult to get at because they burrow inside the leaves, causing blistering and stunting. Spray quassia solution on to the foliage every three weeks in June and July to prevent the flies from laying their eggs. Slugs also like celery and should be watched out for, as the damage they do can form the sites of entry for other diseases, like celery heart rot.

Celeriac

Celeriac, otherwise known as turnip-rooted celery, is another exotic crop that is becoming more popular and is largely supplied by imports from Holland at present. It is hardier in all respects than celery, being more resistant to pests and diseases, standing light frosts and able to be stored through the winter months. It is equally slow growing, needing good growing conditions with perhaps more sun but less moisture. Monarch is a good new variety.

Treatment of the soil can be the same as for celery, deeply cultivated and well manured. Sowing, pricking out and transplanting can also be the same, but plant celeriac as shallow as possible, both when pricking out and transplanting. The swelling root needs to be as much above the surface as possible, otherwise it will be covered in a contorted mass of small and large roots that will considerably increase the trimming time and detract from the shape and appearance of the finished product. Spacing needs to be a little wider at about 30×35cm.

Hoe as required, but irrigation should not be necessary. Lift for storage in October or November, or cover *in situ* for frost protection. If you do get pests or diseases, they will be the same as for celery and treatment can be similar.

Garlic

One crop which has shown a remarkable increase in sales in the last few years, but is still not grown much in this country, is garlic. It therefore has tremendous potential, and all the more so for organic producers as it is a real 'health' crop. Perhaps its main drawback is that seed, that is to say the cloves, are not readily available over here in quantity except as bulbs for ordinary sale from the wholesale markets. To choose and find the variety and quality that you want will therefore require some investigative work around your local market.

It seems that the red-tinged garlic from the mountainous areas of the south and south-west of France are the ones best suited to our growing conditions. The variety Ail Rose d'Auvergne, if your researches get that far, is long established and hardy. Have nothing but the best, and you will require at least 1.5t or about 35,000 bulbs for a hectare's worth of planting. Then settle down for the mammoth operation of breaking up the bulbs into cloves. You will need strong thumb nails or a source of compressed air. Discard any that are small, for the size of the bulb at harvest is directly proportional to the size of the original clove. (You could try growing on the small ones, like spring onions, for pulling green and selling in bunches.)

Garlic likes rich, warm, moist but free draining soil; in fact, it is ideal for raised beds. Given all that, manuring should hardly be necessary, especially if the previous crop was well fed. Phosphate levels should also be good. Make a reasonable tilth, take a weed strike, and plant just below the surface, perhaps a fraction deeper than you would onion sets, at the end of September at 30×10cm spacing. Use irrigation if needed to keep up the soil moisture, particularly as the bulbs swell next spring and summer. Alternatively, it is a crop that benefits from mulching, and black polythene can also be used.

Weeds will need hand hoeing at this spacing but after the winter the stem is thick and strong and the bulbs are surprisingly deep, so there is very little risk of damage. Harvest towards the end of July. Timing is critical, for the outer skin quickly deteriorates if garlic is left in the ground too long. Wet conditions generally at this time will cause similar problems, and rotting will soon take a hold. If the weather is unstable, do not rely on field drying, but gather in the crop and air dry it under cover. For long-term storage, it is essential to dry it very thoroughly.

The main diseases are white rot and eelworm. Both are soil borne and can be introduced on infected stock; if you can get certified seed, so much the better. Once established, you could save the biggest and best bulbs as seed for your next crop.

Jerusalem Artichokes

Jerusalem artichokes have nothing to do with Jerusalem and are not related to the proper globe artichoke. In fact, their foliage and the height they reach gives them away as being related to the sunflower family. There is good sale for this trouble-free vegetable, which can be left in the ground until it is required right up until March. The only way of procuring seed, or rather tubers for planting, is buying in bulk on the open market and selecting the best, most even and smoothest for use. Thereafter make your own selection from your own stock.

The soil for artichokes must be deep and well manured. Up to 50t/ha of farmyard manure can be applied and worked in, but this may have to be done the previous autumn as they need to be planted out as early as possible. Plant in March or even February, some 15cm deep at 30 to 45cm spacing, in rows far enough apart to be ridged up later on in May. Cut the stems off in the autumn and lift as needed. Be sure to lift all the tubers, otherwise you will end up with a volunteer crop next year, unless you have pigs, who love them, to forage for them.

The only possible pest problem you might find is root aphid, which needs to be cleaned up by washing your next year's seed tubers with nicotine. This will kill the eggs as well.

OTHER CROPS

Spinach and Chard

Most organic growers tend to stick with perpetual spinach and Swiss chard rather than launch into the successional sowings of ordinary spinach that are more suited to machine harvesting for processing than the fresh trade. The choice between them rather depends on your market, remembering that there is also a coloured Ruby chard which could add novelty value to your sales. There is no other choice of variety, which rather reflects the lack of breeding that these two types have experienced. Their variable quality bears this out too.

The fertility from a previous crop should be sufficient if the soil has been well manured. This is in marked contrast to ordinary spinach which needs good feeding and rich soil to prevent it from bolting too early. The crop can be sown into well-prepared soil after a weed strike in April and again in July. Alternatively, plants can be raised in blocks and transplanted at those times to give spacings of about 30×35cm. Some sources suggest chard should be at greater, and perpetual at lesser spacings, but the former can easily get too large and the latter make easier picking if larger.

Hoe until the crop can look after itself and start picking when the leaves are large enough, taking the outside ones and leaving the young ones to grow on. Pull them like rhubarb rather than breaking them off, and do not discard the storks. They can be a delicacy in themselves and are, anyway, the heaviest part. Whether you work successionally through the crop or pick lightly over the whole lot every time is up to you, but take care to keep the crop tidy and, when packing, watch your quality control. Oversized, torn, diseased or eaten leaves do not look or taste good. Break off any bolting spears as they appear, the smallest of which can be included for sale.

Pests and diseases should not trouble these two crops very much. Leaf spot will claim some leaves and downy mildew some others, particularly as the weather turns in the autumn. Good housekeeping should suffice but use the normal organic fungal treatments if things get out of hand. As the crop slows right down in the winter, the leaves become more susceptible to damage, so you need to time your last pick to avoid this as much as possible. Perpetual spinach at least will last the winter, and crop again in the spring, for which a tonic feed of dried blood or something similar will help get things going again.

Courgettes

Courgettes were an exotic vegetable some ten years ago and are now staple fare for many small growers. They yield well, are relatively trouble-free under an organic system and need the constant attention, in this case for picking, that the intensive producer can give. There are many varieties, but some are terrible. President probably tops the bill, followed closely by Elite, Ambassador, Sardane and Diamond.

Courgettes like rich, moist soil, so apply up to 75t/ha of farmyard manure and work the soil to a good tilth. The yield from a crop depends on the length of the growing season among other things, so the earlier that cropping can begin the more you will get. Therefore start plants off in modules at the end of April, and transplant in mid to late May, preferably through a black plastic mulch or under cloches. Bare ground is fine, especially in warmer regions; you will lose out in earliness rather than in overall yield. Spacing will depend on the system, but 60×130cm in single rows or 60×75cm in beds, with larger wheelings in between, should be adequate.

Without mulches the weeds will tend to grow through what is apparently impenetrable leaf growth, so hoeing will be needed. Irrigation can also be used to advantage, but is not necessary. Warm weather is more important. Once in full flush you should really pick through the crop every day, for they very soon grow away and will then need to be left to become marrows. These can be a useful bonus as long as you do not have too many of them, for courgette types do not really produce quite the right shape for a true marrow.

The main disease is mildew, which arrives as the weather cools and dampens. Sodium bicarbonate and seaweed spray will take care of any trouble, but it is unlikely to get out of hand. Aphids may cause some problems, principally by spreading cucumber mosaic virus. Normal organic control of the aphids might be necessary, together with removal and burning of the virus-infected plants, but this will only be needed in very few years.

Pumpkins and Squashes

Squashes are one of the 'up and coming' crops at the moment, with a myriad of weird and wonderful shapes, sizes, colours, textures and, for the more discerning, tastes. Our traditional British equivalents, the marrows and pumpkins, make very poor relations to them. Perhaps it is no wonder that they are being superseded, as our national palette is slowly educated out of its philistine ways.

There are summer and winter types. The summer ones tend to have a bush habit and the winter ones trail. Our own types come in the summer bracket, being bushy and only storing until about Christmas. Apart from marrows and pumpkins, there is no choice of varieties, you just choose the type that you want. The choice in pumpkins is more a matter of size, with the smaller Japanese varieties becoming quite popular. Good summer squashes include custard marrows, patty pans and scalloped. Winter ones include buttercups, acorn squashes, turks turbans and hubbard squashes.

All these require very similar growing conditions to courgettes. Soil preparation, propagation, fertility, weed and disease control can be the same. Spacing should be a little more, however, aiming nearer 90×120cm. The problem comes when the trailing ones start trailing, for they are extremely vigorous and sometimes look like they would take over the whole holding, if given the chance. You will probably need to relay their growth in the direction that you want, for instance up and down the rows. Weeding then becomes difficult, so like courgettes they would be a good crop for black polythene mulch.

Asparagus

Asparagus has been on the decline as a crop in this country for many years, but just recently it has begun to make considerable headway again. As an organic crop it has not even started yet. It is difficult to grow, slow to establish and start cropping, labour-intensive and hungry. As a result it is a delicacy of high price, and possibly worth considering, if you have space to play with that is weed-free, warm, rich and well drained.

The soil needs to be well prepared and may need up to 150t/ha of farmyard manure. One of its great disadvantages used to be the method of planting, by crowns, but now everything has changed as modules have hit the asparagus scene. Seedlings, not crowns, are planted, and mechanisation can be employed. These are planted in May or June, after which, treatment is as for crowns.

Crowns should be planted carefully in March or April at about 60×90cm spacings, covered but with a depression left over them and allowed to grow unpicked for preferably the next two years. The new crowns will be very susceptible to dry conditions so this will need watching. Feed with 1t/h of something well balanced like fish blood and bone for the first two years, pick for a month the next year, six weeks the following one and eight weeks thereafter. Cut and remove the ferns in November and mulch with manure.

The main problem is likely to be weeds, for asparagus requires a permanent bed with no competition. Heavy mulching is probably the best answer; plastic mulches cannot really be used. If perennial weeds are present or take a hold, you really will have problems. The asparagus beetle can also be a serious problem, requiring derris and pyrethrum mixed during cropping, and nicotine the rest of the summer. Removal of the ferns at the end of the year will also help.

Herbs

The popularity of herbs for culinary, cosmetic and medicinal uses means that there is great scope for their production, and particularly organically because of their superior quality and purity. Each herb obviously needs different growing conditions and has different problems and advantages. Unfortunately there is not the space here to do more than introduce the idea. Our climate is not really ideal for many, if not most, of the herbs that are commonly used, as they are mainly a sun-loving group of plants. However, not all of them are ruled out, and there is considerable demand for those that will grow well here. There are various ways of preparing, processing and marketing them yourself, if they are to become your speciality.

One word of caution regarding dried herbs: European growers can produce them far better and cheaper than British growers, so do not expect to compete with them. Organic herb production must be much more specialised, such as producing the highest quality of specific types for medicinal use, or producing fresh herbs for the catering and domestic market.

7
Field Vegetables

The main points that characterise field vegetables arise from the scale that is involved; larger areas, sown or planted at one time and harvested all together make the planning of the rotation and of labour requirements a much simpler task. However, there is a need for an equal scale of buildings, perhaps more specialised machinery and the ability to accommodate large fluctuations in workloads. Whilst this might be more possible on the bigger farms, field vegetables can still be worked into the smaller, more intensive systems, and some crops are equally at home there, as a break crop in a standard arable rotation.

PLANTING SYSTEMS

With a few exceptions, most field vegetables make larger plants which require more space. Wheel tracks do not get in the way quite so much and are more easily accommodated between the rows. The bed system is still a major and useful method, but two others can be added that are also appropriate to planting on a field scale.

The first is hardly a system or a method, for it simply involves row crops 'on the flat'. The width of the rows is determined by the tractor wheels, and drilling or planting and subsequent operations can all be done using the same wheel tracks. The normal tractor widths are 140cm, 150cm and 160cm, which would give standard row widths of half of those figures, or a third of them if three rows were to fit between the wheels. This would only really be possible with the wider wheel spacing, otherwise there would hardly be enough room for the wheels to pass through the crop without damaging it.

The second method is almost a variation on this theme and involves ridges. Only two ridges can fit between the tractor wheels, and they can be formed either whilst the crop is being planted, as with potatoes, or beforehand so that drilling is done on the ridge, or after the crop has established itself, as with leeks. Ridging can be considered as a growing system or simply as a cultural technique to be used in conjunction with row crops. It is perhaps more widely used by organic growers because of its advantages with weed control and the fact that the mini raised bed, formed by the ridge, encourages good biological activity. Conventionally this system is confined to the wetter western parts of Britain, apart, of course, from its widespread use with potatoes.

LEGUMES

Peas

Peas are rather an in-between crop, because they are grown on a vast scale in eastern England, mainly for processing, and also in much smaller amounts for very laborious hand picking for the fresh market. I am not sure how the price ever justifies this latter method. Furthermore, the timing of harvest is critical for the flavour and sweetness of the crop, so it nearly always suffers as a result of hand picking. Fresh peas are often mealy and tired, and it is easy to see why the processors demand such tight schedules of their vining peas.

One recent development has revolutionised the pea crop for the small grower however, and that is the sugar pea. Concentrate on this, of which there are two types: the true mange-tout that is picked when the pods are still flat, and the snap pea from the United States that can be harvested when the pods are swollen. Both, but particularly the snap pea, are increasingly popular, with good prices being paid and demand only just beginning.

The seed catalogues do not always make it clear which type is which, so be careful when choosing, and be aware of the height of the haulm, to know whether it needs staking or not. Try the mange-tout variety Caroubel, snap peas Sugarbon (early) and Sugarrae (maincrop), and the ordinary Feltham First for autumn sowing, and Progress No. 9, Kelvedon Wonder (for particular flavour) and Hurst Greenshaft successionally for the spring and summer.

Peas like reasonable soil with good potash levels and neutral pH, but manure can be considered a luxury for them. They do like good drainage and cultivations must take this into account. Otherwise make the seedbed reasonably firm and not too fine. Allow for a weed strike, as weeds can be the bane of peas. Sow from March onwards, on anywhere from 25 to 60cm rows at about 1 to 2cm spacing, depending on which system you wish to use: in the narrower rows the peas will support each other better, but the wider rows can be ridged.

When the peas emerge, and again when they are about 6cm tall, go over them with chain harrows or something similarly shallow for further weed control. After that, they should remain fairly clean, but you can hand or wheel hoe, or even steerage hoe on the wider spacing, and then ridge before the time comes when you cannot get through any more. For harvest it is best to pull the whole plant and pick everything from it. The crop can even be cut and carried to be picked elsewhere. The main point is to ensure that it has not gone over, but equally that it is sufficiently mature to give the yield. The only pest, besides pigeons and rabbits, that you may encounter, is the pea moth larva eating away inside the pods. There is no treatment, but good rotations should keep it at acceptable levels.

BRASSICAE

Winter Cabbage

Winter cabbage includes white, January King types and savoys, and the new hybrids between them. There is always a good demand for them, and choice of the right varieties will give you cabbage to cut right through the winter. One area that is totally neglected in Britain and is wide open for exploitation is the storage of organic white cabbage. Massive amounts are imported at present, but the quality is superb and any substitution from British crops will have to be equally good.

It is difficult to choose between varieties, but table 7.1 makes some brief suggestions and includes some basic information.

All three white cabbages are excellent for storage as well as for the fresh market. Sow them at the end of April for planting out in early June and harvesting from November onwards. All the others need sowing in May for transplanting at the end of June and cutting from December to March and even April, depending on type, variety and growing conditions. A later sowing of savoys for planting up to the end of July (in the south of England) can be made to ensure the viability of this later cutting.

Although bare-rooted transplants are adequate, the shock and failure rate can be substantial, especially in dry conditions and with dry soil following an earlier spring crop. Modules or peat blocks may therefore be well worth using, to ensure strong, even plants and to achieve a good take without having to use irrigation, which would otherwise be unnecessary.

Winter Cabbage Varieties

Variety	Planting–Maturity	Average Weight	Standing Ability	Density
White				
Slawdena	22 weeks	1.35kg	fair	extremely dense
Bartolo	24	1.6	good	extremely dense
Marathon	26	1.6	excellent	very dense
January King				
Aquarius	21	0.79	fair	fair
Marabel	23	0.86	excellent	fair
Savoy × White Hybrids				
Corsair	18	1.55	excellent	dense
Celtic	20	1.49	extremely good	dense
Pict	23	1.42	very good	dense
Savoy				
Iceprince	21	1	very good	dense
Tarvoy	25	0.84	extremely good	very dense

Table 7.1

Red cabbage for storing can be treated just like white cabbage, which it resembles closely, apart from the colour. Kwanta is a good variety for this.

Winter cabbage, like others, is a heavy feeder, so soil preparation should include a generous spreading of farmyard manure, up to 50t/ha, and provision for correcting pH and potash levels if these are low. Cultivate to produce a deep but firm tilth, with minimum passes to conserve moisture, and plant at any spacings from 40×40cm to 40×70cm, depending on your system and the size of heads that your outlets prefer. Generally the move is towards smaller cabbages, so it is perfectly possible to plant them quite tightly. Weed control might also determine the spacing; winter cabbages can be ridged as they grow, if planted on appropriate row widths.

Pests and diseases are similar to those affecting summer and autumn varieties. One aspect is different, and that is the cold weather. White cabbages are not frost hardy, so do not expect them to last too long in the field. Many are lost every year because they should have been cut and stored before the weather got too severe. Storage is not difficult, although cold stores with controlled atmospheres are needed for consistent, long-term results after April. Up to the end of March, clamps or barns can be successfully used as long as the cabbages going in to store are of the highest quality. Trim the loose outer leaves and handle with care to avoid bruising and abrasion, which might lead to rotting. The ideal conditions are a temperature of 0 to 1 degree Centigrade and a relative humidity of 95 per cent.

Spring Cabbage

If you can keep the pigeons and rabbits off spring cabbage, they are a welcome crop at a time of the year when little else green is around and the farm account is usually in need of a boost. They are versatile, for you can almost double crop them – half as leafy greens, and the remainder as hearted cabbage. They can also make a good catch crop, being in the ground for an unusual season, that might help to make full use of a rotation's cropping. However, do not underestimate their cultivation. They are a tricky crop for organic production as they need to be growing in the late winter, when the soil life is not really active.

First Early Market 218 is probably the best early variety, more suited to greens than hearted. Myatts Offenham Compacta, a little later, will do both. Spring Hero is the only overwintering ball-headed cabbage which might be worth a place in the cropping programme. Bare-rooted transplants are fine for spring cabbage, but direct drilling in the field can be better, particularly for the early greens crop. In order to achieve good continuity, a mixture of varieties and methods could be used, for example direct drilling Early Market in early August for spring greens, and sowing Compacta at the same time in seedbeds for transplanting in late September. With appropriate earlier sowings, the cropping period can be extended back into the autumn as well. Final spacings want to be close, with rows down to 35cm apart and plants even down to 10cm apart, if every other cabbage can be cut for greens, to allow the rest to heart up.

The fertility in the soil from the previous crop should be sufficient to start with in the autumn. The problem comes in the new year. Feeding really has to start in January for the early greens, to allow the nutrients to work through the system. Dried blood is probably the only one that is quick enough and can be used at up to 700kg/ha then, and probably again in March for the later ones. Try to place it close to the rows rather than scattering it, as it is expensive.

Cauliflowers

The cauliflower is supreme amongst vegetables. It will crop throughout the year and is always in demand, but it requires the best of growing conditions and management. The number of varieties and types is extremely bewildering, and is only briefly outlined here. The earliest in the summer and the latest in the autumn and winter are the poorest in quality, but command the best prices. Some of them will only grow in the mildest southern areas, so check carefully before embarking on anything too ambitious.

The best soil is fairly strong, fertile, free draining and with a pH of at least 6.5. Give a good dressing of up to about 50t/ha of manure and cultivate well. The later ones, going in in mid-summer, may require less if the previous crop was well fed, or alternatively a general fertiliser, fish blood and bone for example, at 700kg/ha can be used. Modules or peat blocks are by far the best way of propagation as cauliflowers do not like

Cauliflower Types, Varieties and Cultural Hints

Variety	Sow	Transplant	Transplanting–Cut	Quality
Early summer				
Mechelse-Panda	early Oct. or	Mar.–Apr.	14 weeks	medium
Corvilia	Jan., with		15	good
White Summer	heat		16½	very good
Late summer				
Andes	Mar.	late May	11	very good
Early autumn				
Andes	Apr./May	mid–end June	10	very good
Late autumn				
White Fox	mid May	June/July	10½	very good
Snowy River			16	medium
Winter (south west)				
Alor	late May	July	23	good
Briac			32	good
Spring (Walcheron Winter types)				
Armado April	May/June	July	39	poor
Asmer Bostonian			41	medium
Matterhorn			44	medium

Table 7.2

delays and checks. They are quite slow but they have a lot of growing to do and need to be healthy and strong all the way through. Spacing should be around 60cm^2, giving the winter and spring varieties slightly more space and the summer ones slightly less.

Give the summer crop a tonic feed of dried blood at about 400kg/ha once it is established. Irrigation can be helpful in the summer too, if conditions get dry, especially some three weeks before maturity. Once they are ready, pick young and regularly, while the curds are still well protected for they do not keep in the field. Handle them with great care for the curds are very sensitive to bruising and damage.

Pests and diseases are the same as for other brassicae, including cabbage root fly, clubroot, caterpillars and aphids. Control or prevention are also the same.

Calabrese and Broccoli

Although these two are closely related, they are really winter and summer versions. The greatest breeding has gone into calabrese which now sports a welter of F_1 hybrids and has become a very popular choice vegetable. Sprouting broccoli is not very high yielding but its virtue is that it crops in the lean times of early spring. Try the calabrese varieties SG1, Premium Crop and Shogun for early, mid-season, and late. There are only Early and Late purple sprouting broccoli varieties. Two other more unusual types may be worth considering for the specialist market: Purple Cape and Italian Autumn broccoli.

Calabrese and Italian Autumn grow quickly and need the same soil treatment as cauliflowers, including the use of modules. Sow calabrese any time from February (with heat) until June, for planting out at 30×125cm and cropping throughout the summer and autumn. Purple sprouting and Purple Cape broccoli need to be hard and strong for overwintering. They do not like a rich soil and can rely on a previous crop's manuring. They can also be bare-root transplanted. Sow purple sprouting in April for planting at 60×70cm in early June and harvesting the following February to May. Italian Autumn and Purple Cape broccoli can be sown a month earlier, for cropping up to the first heavy frosts, and in February and March, respectively; spacing is at 60×60cm.

Irrigation will help the calabrese and Italian Autumn, but is not necessary for the overwintered ones. Calabrese will produce side shoots after the main head has been cut, if you can justify the labour of picking them. With all types it is important to keep cutting the spears in order to maintain production. Treat pests and diseases as for other brassicae.

Brussels Sprouts

Sprouts are not the most profitable of crops, but their organic production is hopelessly below the demand and it makes one of the most suitable of field crops, being vigorous and weed-suppressing when established, and requiring work at times that are otherwise slack. Varieties can be chosen that will spread the cropping from late September until March. Try

Oliver, Roger, Golfer and Gabion for a good succession.

Modules are not so important for this crop. The earliest two can be sown with protection in March for planting out in May, and the later two can be sown in an outside seedbed in April for transplanting in June. Final spacing can be anywhere from 45–60×70cm. They like firm soil in which to anchor, and it should be fertile but not rich, so manuring is not necessary if the previous crop was manured. Watch the pH, as with the other brassicae.

Irrigation will not be necessary, except perhaps to help the young transplants establish. As the crop grows it can be ridged, both for weed control and to give it support against the winter weather. You should manage up to three picks during its cropping life and, if the pigeons do not interfere, which they almost inevitably will, you may even be able to harvest and sell the tops, although this may be at the expense of the final picking. Pests and diseases are the ones normally affecting brassicae.

Swedes

Swedes are a stalwart of many a farm rotation. It is one organic crop that is almost well supplied, so do not embark on too large an acreage without first finding an outlet for it. The supermarkets are taking large quantities, but their quality control is very strict. When you have seen a Devon swede, grown on wonderful red soil, beside any other, you will have second thoughts about growing them outside Devon. Marian is probably the best variety.

As with other brassicae, swedes like their soil fertile, well drained and fairly firm, but slightly more acid at about pH 6 to 6.5. Manure either with a light dressing of 20t/ha or not at all, if the previous crop was well fed. Cultivate to a good depth, but the seedbed does not need to be too fine. Light soils and many areas in the wetter west of Britain may be low in boron, to which swedes are very sensitive. Borax at 8kg/ha should remedy this. Wetter soils can be ridged to help with earliness and weed control.

Precision drill in June after a weed strike, or in May or even April if you are going for an early crop, in which case floating film can also be used. Aim at about 12cm spacing in rows 45cm apart, or an equivalent density. Ridges will obviously be further apart. Irrigation should only be necessary in the drier parts of Britain, where you probably should not be growing swedes anyway. They have a tendency to get rather tough if they do not keep moving. Harvest is from September to May, but a really hard frost will put an early stop to it. Swedes can be clamped, but this is not normal, although perhaps it ought to be more common practice.

Control the weeds with at least one hand hoeing in the rows, which can be combined with singling if you did not precision drill, and several inter-row cultivations. Flea beetle can be an absolute menace for the emerging crop in some years. Derris pyrethrum, mixed, will then be necessary, or even redrilling if it is really bad. Cabbage root fly may also cause major problems because there is very little to be done about it and the saleable quality of the crop can be badly affected. If you are in a bad area for root fly, do not tempt fate.

Turnips

Turnips are the lesser cousins of swedes, and fewer are grown. However, they gain at the beginning of the season as an intensive crop, and turnip tops can also be sold. They are quicker growing, earlier and therefore more versatile. For the maincrop, choose a green top variety like Manchester Market. Others with different skin and flesh colours might be appropriate in some circumstances. The new white hybrid, Tokyo Cross, can also be used for the maincrop.

Soil requirements are much the same as for swedes, firm and fertile but not overfed. A light dressing of a general fertiliser may be needed after a heavy-feeding, previous crop. Drill turnips in July or August, aiming for a 12×40cm spacing. Ridging is not necessary but a bed system can be used. This late start means that they can be a useful catch crop after an earlier crop, and can also double as stock feed if the market is not there. Irrigation would be needed to get them moving in dry conditions, otherwise they should be fairly free from problems. Flea beetle is not a problem so late in the summer and they might miss the last root fly generation.

Turnips are not really hardy and need to be clamped if they are to last beyond Christmas. Lift them around November and store in low clamps about one metre high, well protected from frost. Do not expect them to keep their condition too long into the new year.

ROOTS

Potatoes

Potatoes need no introduction. They are a staple of both intensive and field production. They make a good cleaning crop and there is a method and a market to fit most situations. Varieties are divided into early, second early and maincrop. Polythene mulches can bring forward the early crop still more and for those growing at altitudes there is the undeveloped option of producing organic seed potatoes. Ulster Sceptre and Maris Bard are still about the best all-round earlies, Estima and Wilja for second earlies, and Cara and Desiree for maincrop. Others might be appropriate for special conditions or markets.

Production should begin well before planting with chitting of the seed. The longer and slower this process is, the stronger and sturdier will be the subsequent growth. If you can get the seed, start off earlies and second earlies in November, under cover first and then inside where it is frost-free. A later start can be compensated for by bringing the temperature up to around 15 degrees Centigrade for the first ten days, to break dormancy, and then keeping it at about 8 degrees Centigrade thereafter. The maincrop can wait until early spring, and can be started off with a week at 15 degrees Centigrade.

Potatoes like plentiful organic matter, and need a deep, loose and friable soil. Apply up to 50t/ha of manure, preferably in the spring and work in with the seedbed cultivations which are aiming to produce sufficient depth of good tilth to form full-sized ridges. Rock potash can also be beneficial. For earlies and less amenable soils, you may have to plough in the early winter in order to have sufficient structure in time. Plant earlies from January (in the most favoured areas) to March, and the others in early April. 2 to 3t/ha of seed, depending on whether certified or once-grown seed is used, will give the right density, the spacing being determined by your tractor wheels.

Weed control can start ten days after planting, using chain harrows and ridgers to knock down and build up the ridges. After emergence, inter-row work can continue, with a final earthing up just before you are no longer able to get through. Minimise the compaction during all this work by cultivating the wheelings. Irrigation will increase yield, but must be kept up once started. Earlies can benefit all through their growth, but the others can wait until the tubers are at the small marble stage. Alternatively, maximum benefit will be gained from irrigating two weeks before harvest.

Potato blight will inevitably strike the crop at some time during the summer and it is then downhill all the way. Regular seaweed and waterglass sprays every ten days from the first blight warnings will help stave off the attack, but cannot be expected to do more than that. Burgundy mixture (10kg copper sulphate and 12½kg sodium bicarbonate in 1,000l of water) is more effective but should only be used with great discretion. Once the disease takes hold it is best to flail off the haulm and leave for three weeks before attempting to harvest.

There are many other diseases and pests. Most are not really a problem, except in unusual circumstances. Common scab likes lime so ensure that any liming is done only after potatoes. Viruses should be avoided by buying good seed, and it is highly unlikely that you will need to spray against their aphid vectors. Nematodes need long rotations, resistant varieties and mustard green manures. Wireworm should only cause damage after permanent pasture; two crops of mustard green manure should help.

Storage is better inside than in a clamp. Go up to 2.5m deep, as more will need extra ventilation, and provide a cool, dark, ventilated, and frost-free environment. If you wish to grow more than half a hectare, you must register with the Potato Marketing Board and apply for a 'quota', which you may not get.

Onions

A fine crop for organic cultivation, onions are always in demand and are still not well enough supplied. Besides the normal maincrop, there are the overwintering Japanese onions. These are more difficult and should only be attempted in milder areas. Technology has advanced greatly with onion growing and now the only method, particularly for the organic producer, is with multiblocks. Try Prospero for yield or Mercato for extra storage. Buffalo is the only one for overwintering. All three are hybrids.

Unless you are an experienced plant raiser, you will be better off buying your blocks from someone who is. There are a few organic ones available. The blocks, sown in January with heat at a density of six or seven seeds per block, will be ready for transplanting in the middle of April. The soil conditions are less critical than for the direct-drilled crop, but should still be reasonably fine and firm. Work in up to 50t/ha of well-rotted manure at the start of your cultivations. Plant at about 25×35cm, aiming for a population of 65,000 plants/ha.

The overwintered onions need sowing in the second half of August, for planting at the end of September. Timing is critical, in order to avoid them bolting in the spring, and advice should be sought for your local area, if you can get it, as this is not common practice in conventional growing. In March they will need a boost to get them growing before the soil has warmed up. Dried blood at 500kg/ha should suffice.

Onions will not suppress the weeds at any time, so weed control is important throughout. Inter-row work will be possible to begin with, and hand hoeing will probably be needed at least twice. Irrigation will always help, especially for establishment, stopping when the bulbs begin to ripen. However, it is not essential. Of pests and diseases, neck rot is seed borne and should not be a problem. White rot is soil borne and at least a six year rotation is needed to overcome it. Both will seriously affect the stored crop. Onion fly and stem eelworm are the two major pests. They are unlikely to be significant, but very little can be done about them. Long rotations will help prevent the latter.

Start to lift the crop when half the onions have gone over,

windrowing them in the field to wilt for ten days. Undercutting will help the operation. They will then be ready for proper drying in store. If conditions are wet during this process, take great care, for the skins soon discolour, reducing quality and increasing preparation time for selling. Long-term storage of onions is a science in itself. Hardly a single organic farmer does it, so the opportunities are wide open. The overwintered crop will not store, but can be prepared and sold after field drying, which will be that much more effective in June and July.

Maincrop Carrots

The field scale required for maincrop carrots brings slightly different problems from the other types. For weed control purposes, the maincrop is usually grown on ridges and the longer, more exploitative varieties are best. Try a Berlicum or Autumn King type, unless you have heavier land when the shorter ones like Nantes or Chantenay will be best.

The soil should be on the light side, fertile and manured the previous season. Phosphorus and Potassium levels need to be reasonable. Start cultivations in plenty of time to allow for a false seedbed, preferably before ridging and again on the ridges before sowing. Drill into the slightly weedy ground at the end of May, when the ground is well warmed and after the first generation of carrot root fly have passed. Give a pre-emergence flame weed, probably about ten days later. Spacing on the ridges should be every 3cm. If you are feeling ambitious you could sow in a band some 6cm wide, achieving a density of about 75 plants per metre in each row. Alternatively, you could try a double row, about 6cm apart on the ridge. Use a precision drill to get the most even stand and to avoid thinning. On a bed system, rows down to about 30cm apart can be used.

Keep the weeds under control by inter-row cultivations, ridging and hand hoeing. It is not until late summer that the carrot crop will have anything like a smothering action. Irrigation should not be necessary.

Harvest comes in two forms. To be sure of avoiding the last carrot fly generation, lift in early October (in dry conditions) and clamp, or put into controlled temperature storage. Alternatively carrots can be stored in the field, protected by soil. Earth up in early autumn, only possible with rows of 50cm or more, ensuring that you have a covering of at least 15cm of soil. This may be difficult if the carrots are already on ridges.

Parsnips

Related to carrots, parsnips are a popular winter vegetable that can easily overwinter in the field and can be sold from the end of summer to the middle of spring. The supermarkets prefer smaller roots, about five to the kilogram, whereas the fresh trade will accept them up to double that size, but huge ones ought to be avoided. Try Gladiator, an F_1, or Alba for prepacking.

Parsnips like lighter soils, deep and manured for a previous crop. They might benefit from rock potash, but otherwise are not demanding. The same arguments apply about ridging as for carrots. Create a deep, fine seedbed and follow the same weed control pattern as for carrots, because parsnips, like carrots, are slow and small to start and must not face competition. Sow, preferably with a precision drill, any time from early spring onwards, but mid to late April is fine for all but the earliest crops. Pelleted seed can be used to improve the precision of sowing, although research suggests that natural seed establishes better. Drill at 5cm spacings on ridges or at 9×40cm on the flat for the fresh market, or at 6×40cm for prepacking: the right density for prepacking would be difficult to achieve on ridges.

Weed control in the established crop is the same as for carrots. Similarly, irrigation should not be necessary. The only disease of note is canker. There is no control other than using resistant varieties. The main pest is carrot root fly. Again, nothing can be done except ensuring that parsnips and carrots are not grown close together and that they are far apart in any rotation. Parsnips do not store very well and are best left in the ground to be harvested as required.

Leeks

A staple of many rotations, both intensive and field, leeks are ideal to fit in after an early crop has been harvested. Apart from the labour of planting, their work mainly centres around lifting during the autumn and winter, but even then it is spread out over a very long period of time. The leek crop is therefore versatile, trouble-free and popular. There are three basic groups that harvest from September to December, from Christmas to March and in April and May. Try King Richard for the earliest and longest, followed by Jolant in the first group, then Autumn Mammoth, Goliath and Snowstar, and finally Kajak and Carina in the last group.

Leeks can be direct drilled, but not as an organic crop. Instead, transplanting either bare rooted or in multi-blocks is best, depending on what suits your market and your system. The latter go in earlier, at greater density and smaller size, producing smaller leeks of lesser quality. Ridges or beds can be used. Ridges will tend to produce a lower yield of longer, blanched, larger leeks. Dibbing will produce the best and longest leeks, and the process can now be aided by mechanisation.

Sow, not too thickly, in seedbeds in February or March, taking great care about weeding, with false seedbeds and pre-emergence burning, for the seedlings are slow and tiny. The earlier ones need to be inside, and the later ones will benefit from some protection. The standard problem is having transplants that are too small and too late, so that most of your crop starts to come ready when the weather warms up in April, and you are rushing to clear them before they bolt, along with everyone else. With multi-blocks, aim for about four plants in each, sown at the same time.

Leeks like rich soil, so give it a generous 50t/ha of well-rotted manure, and cultivate to produce a good deep tilth. If going in after an earlier crop that was well fed, then something like fish blood and bone at around 750kg/ha will suffice. Plant blocks in May and June and bare-rooted plants, preferably pencil thick and with excess leaf growth cut off, in June and July. Plant on the flat – any ridging is done later. Spacing on ridges needs to be as short as possible, down to 8cm. In beds, use 30cm rows with plants 10cm apart. Blocks go in at double the spacing in the row on both systems.

Irrigation is almost essential to establish the blocks, and is certainly beneficial for bare-rooted plants. After that they

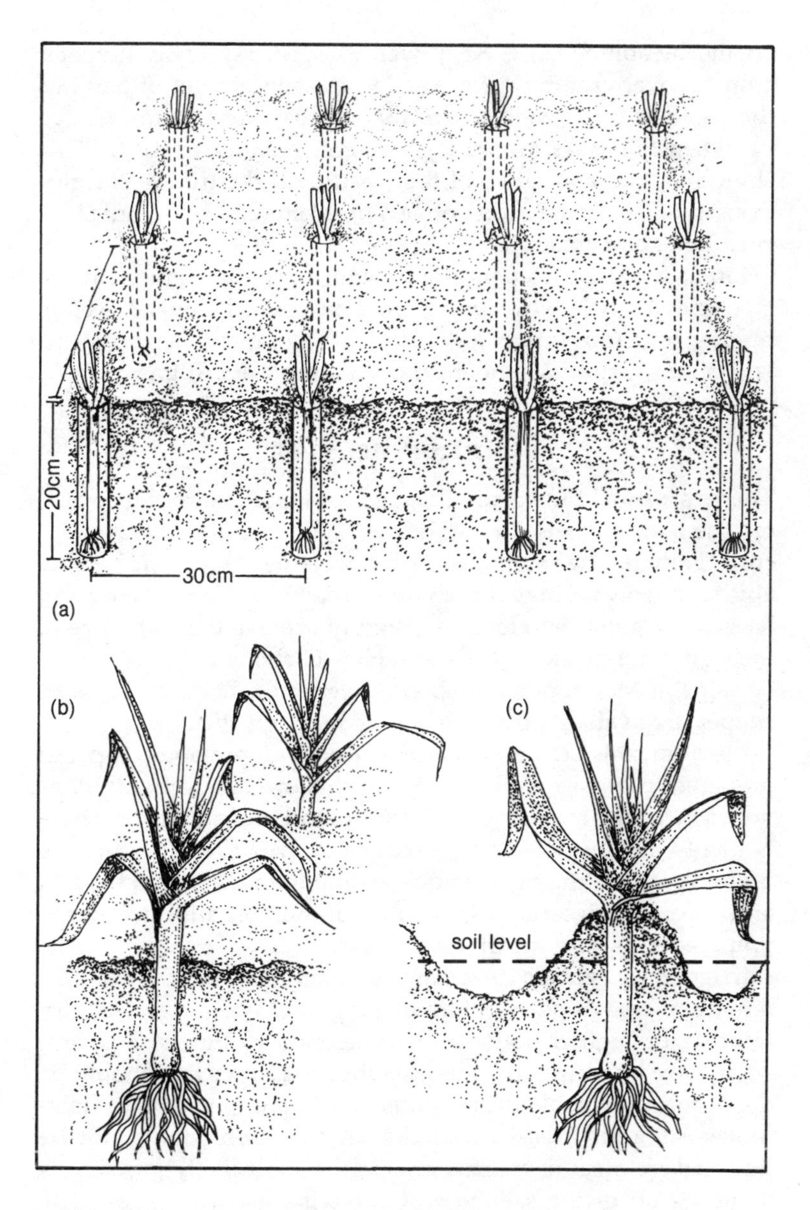

Methods of planting leeks: (a) dibber holes, made by machine or by hand, with young leeks just planted; (b) leeks well established and ready for ridging; (c) ridging completed.

should be able to cope. Keep well weeded, especially the later crop for overwintering, for weeds can soon get out of hand as the weather turns wetter with the approach of autumn. Ridge up when the leeks are strong enough to stand up to it. This should take care of much of the weeding if the timing is right. Pests and diseases should not be a problem with an organic leek crop in a good rotation.

Lift as required. Depending on your scale there are plenty of machines to help you, but if you have the labour, selective pulling by hand with a fork, worked into an efficient system, will certainly produce the most even sample and highest yield.

Maincrop Beetroot

There is a small but steady demand for loose beetroot throughout the winter. It should not, however, be considered as a major crop. Perhaps the main consideration is its size, for it has the ability to grow far bigger than the market likes, and sowing and harvesting must therefore be timed to achieve the right size at the right time for storage. Detroit New Globe or Boltardy, or the monogerm Monogram are all good varieties. Beetroot in other shapes are available if you have a market for them.

Medium soils are best, manured well for a previous crop and prepared to a reasonable but fairly shallow tilth. Sowing does not start for the maincrop until the second half of May, so there is plenty of time for a false seedbed or two. Density must be high to produce the right grades, so aim for a spacing in the bed of 3×30cm. Ridges are obviously not suitable unless you are prepared to have a much lower yield.

Irrigation might be necessary to ensure good and even germination, best applied before drilling, and later on in hot, dry weather on lighter soils. Hoe as necessary, but with a pre-emergence burn and the other seedbed measures this should be down to manageable proportions. There are long lists of beetroot pests and diseases, but the organic crop should not be troubled by most of them. Perhaps the most likely problem is damping off of the seedlings. Otherwise, beetroot is a hardy and vigorous plant, usually able to grow away from any setback it may encounter.

Lift the crop when you have the required size range, prefer-

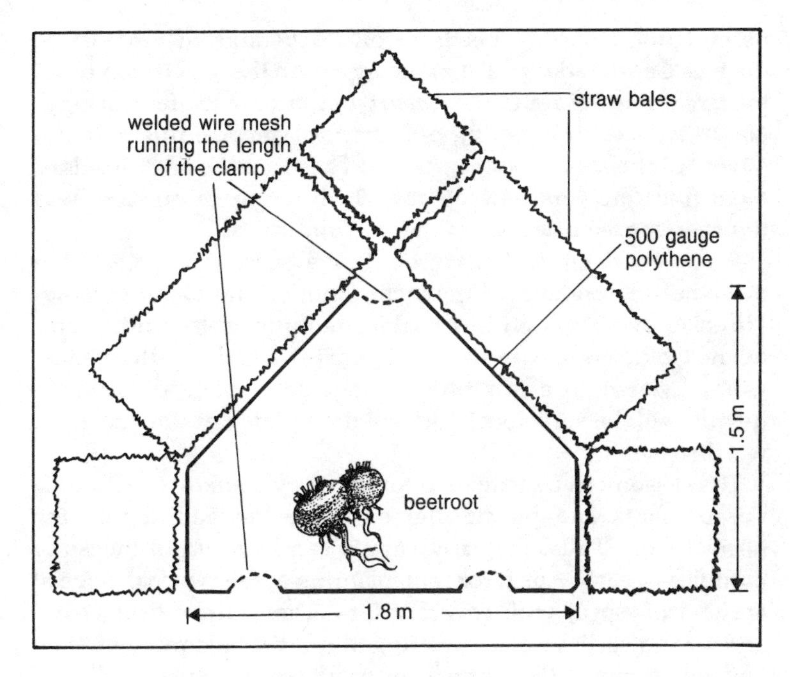

Beetroot clamp, using straw bales and a polythene sheet.

ably not too early in the autumn when it is still warm, but by the middle of November, to avoid frost damage. Store either in a traditional outdoor clamp or in one constructed of straw bales, lined with polythene and with mesh air ducts running its length. The two characteristics of stored beet are that it heats up more than most and that it desiccates very easily; both must be avoided to maintain quality.

OTHER CROPS

Sweetcorn

Sweetcorn is really restricted to the southern half or even third of Great Britain, as it requires warm temperatures and a long growing season to produce well. The most critical times are at germination, when it needs a soil temperature of at least 10 degrees Centigrade, and at maturity, when the sugar soon turns to starch. Furthermore, the quality does not hold after picking. New advances in breeding have partly overcome this with the advent of the semi-sweet varieties. These are the next development from the super-sweets, of which the main problem was the even more critical germination conditions.

For a good succession through the season, try Kelvedon Sweetheart, Sundance, Comanche, Jubilee and October Gold. The semi-sweet varieties available are Butterscotch and Champagne Gold. Sweetcorn like deep, well-drained, medium soils, worked down to a fair tilth. A moderate 25t/ha dressing of manure will be sufficient. You will have plenty of time to take a weed strike.

The season can be extended forwards by a mixture of module raised plants and plastic film coverings, in addition to the normal crop. Thus, the early variety can be sown in blocks or modules in early April, for transplanting at the two leaf stage at the end of April, with covering film. Then, direct drill earlies and second earlies with a covering film at the beginning of May, and finally make the normal, unprotected sowings of all varieties in mid and late May. Spacing should be around 20×75cm for earlies and 25×75cm for the others.

Irrigation is helpful in drier areas, and most critically at the tasselling stage. Weeds should be kept under control with steerage hoeing and probably one hand hoeing, or the crop can be ridged, while it is still small enough to go under the tractor. Frit flies are the main pest, but only of the drilled and unprotected crop. There is no organic control but you are unlikely to need one. Aphids in between the husks can sometimes make the ears unsightly and even unsaleable. This is really an overspill from the rest of the plant, which can only be prevented by regular spraying, but the organic crop should not be infested enough to produce such a bad situation.

8
Fruit

Soft fruit in particular may have an important part to play in the overall balance of a small farm. The returns can be substantial, but the labour requirements tend to be highly concentrated and often in busy times of the year. The picking season, for instance, is usually short and very labour-intensive. Furthermore, the fruit does not keep and needs to be moved on quickly. Alternatively, soft fruit crops are ideal candidates for processing, either into jams and preserves or into fruit yoghurts, in which case ample provision for freezing should be available to provide supplies throughout the year. The red, soft fruit are essential crops if you are considering a pick-your-own enterprise, as they are the real crowd pullers, not to mention the real money spinners.

The top fruit situation is more difficult to assess. It is unlikely that the costs of establishing an orchard could be borne by a small farm, as it would be some years before returns really started coming in. This may partly explain why so little organic top fruit is grown in this country. The other reason is that they are difficult to grow organically to a commercial standard, and it is that much more difficult when there are so few examples to follow. The commercial top fruit industry is extremely chemical-intensive, with all manner of sprays to regulate every stage of growth. The organic alternative cannot match that degree of regulation and, hence, uniformity. There is some comfort though, for there are considerable areas of organic top fruit in Europe and the aspiring top fruit grower would do well to go and investigate them before embarking on anything too ambitious. At least the market is wide open.

With all fruit, remember that the plants will be in the ground for several, or even many, years. Therefore do not skimp on your original plants, or indeed on your planting preparations. Ministry certified, quality stock of carefully chosen varieties is an essential, for although the initial outlay on the best may be higher, the eventual returns will repay that extra investment many times over.

SOFT FRUIT

Strawberries

Strawberries are the king of soft fruit. Much breeding is still going into them in order to produce continuous cropping from different varieties and systems right through the summer and well into the autumn. The tunnel crop produces the earliest pickings; next come those under cloches, then the open air crop and then those either deep-strawed or mulched with white-on-black polythene, to slow down the rise in soil temperature. From mid August the 'ever-bearing' remontant varieties start cropping. In addition to those suggested for the protected crop, you could also try Gorella (early), Cambridge Favourite (mid), Bogota (late) and Rapella (ever-bearer), but do not expect the same flavour.

The soil needs to be rich in organic matter, fairly acid and well drained. Before planting spread 1t/ha of rock potash and 120t/ha of well-rotted farmyard manure, more if the land is light or chalky, for remember that strawberries may be in the ground for up to four years. Probably the best system for organic production is on beds mulched with black polythene, as weeds are the biggest problem in this perennial crop. If you can create raised beds, so much the better. Plant in August for cropping the next summer. The ever-bearers can go in later, even up to the next May, for cropping that year.

Spacing depends on your system and the varieties used, but do not go below 30×45cm and remember to allow plenty of room for access between the beds. Tractor work will need more, as will pick-your-own pickers. If you prefer something more intensive, you could plant three rows to a 90cm bed and grub out the middle row after the first season to give 45cm rows. In this way, they can be cloched in the first year for an early crop and treated normally thereafter.

Weeds should not be a problem if black plastic mulch is used, although make sure the area is clear of perennial weeds before planting. Weeds in the pathways will still need to be controlled however, by hoeing or mulching with straw or other material. Trickle irrigation under the polythene is an advantage and probably necessary in the drier east. It is possible to do without as long as the soil is at field capacity when the plastic mulch goes on. A compromise, though more expensive, is to use Mypex, a woven black polythene. It is thicker and stronger but allows a certain amount of moisture to penetrate. It might also be considered for a pick-your-own crop, where wear and tear is likely to be heavier.

Aphids are the main pest which, if large numbers build up, can be controlled with a good covering of derris-pyrethrum mix in April, repeated again five days later. Mildew and botrytis are the main diseases. Lime sulphur at least three weeks before picking is effective for mildew, but good housekeeping is the best remedy for botrytis. Ensure that the plants do not get too crowded or weedy to maintain ventilation; remove rotting fruit and clear runners and old vegetation after fruiting has finished.

Raspberries

This is the next most popular soft fruit and it too has been the subject of intense breeding to prolong its cropping season. It does not lend itself to different cropping systems as the strawberry crop does, except for minor differences in pruning techniques, so the length of the season entirely depends on choosing the right varieties. A good succession could be achieved from early July to October using Glen Moy, Malling Jewel, Malling Leo and the new autumn fruiting variety Malling Autumn Bliss.

It is essential that your chosen ground is clear of perennial weeds before you start, because once the raspberries are in, you will not get another chance of the clean sweep that their elimination requires. If there is a problem, allocate the last half of the summer and the autumn to pull the weed roots to the surface to wither and die. Use spring tined cultivators or something similar, going over the ground regularly, and gradually increasing the depth each time. At a point when the soil is dry during all this, give the ground a good subsoiling.

Raspberries prefer their soil to be rich in organic matter, well drained and slightly acid (pH6), much like strawberries. However, they will be in the ground three or four times as long, so need even more manure to be incorporated before they are planted 8cm deep, preferably in November. Manure at 200t/ha would by no means be too much, along with the same 1t/ha of rock potash. Repeat the potash dressing every four years, and spray annually with liquid seaweed, after the buds break and again as the fruits form.

Spacing is determined by whether you need tractor access between the rows or not. The minimum is 180cm rows and 40cm between plants. Cut the planted canes to about 25cm so that they do not flower and fruit in their first season. Thereafter, pruning should be as soon as possible after fruiting has finished. The autumn varieties need all their growth cutting right down in February.

Now comes the difficult aspect which is common to all soft

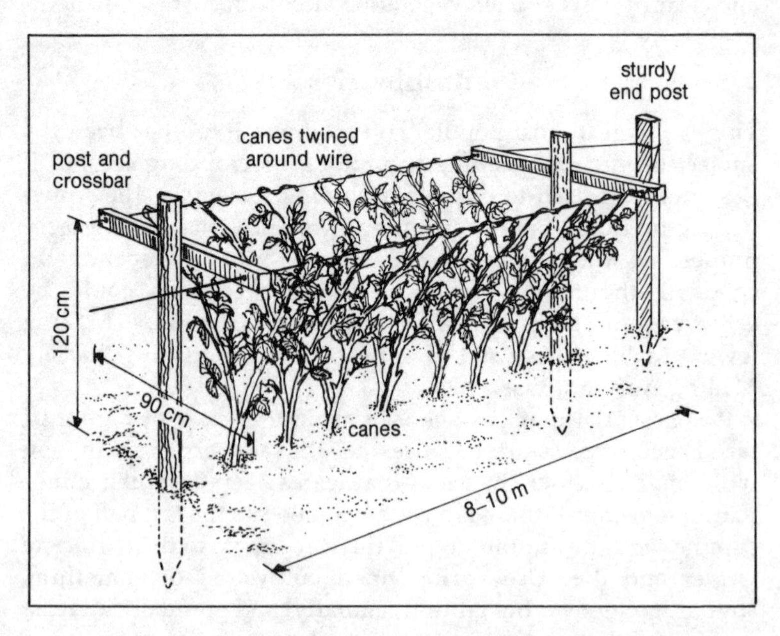

Raspberry training; the canes are trained in both directions and twined around the wire.

fruit: weed control. You have three options for short-term weed control. The first is to mulch with black polythene, which you would plant through, and which would give control for at least the first season. It is not a long-term measure but will give you a clean start, and that is important. The second is to mulch deeply with a good 15 to 20cm of straw, wood chippings, bark or other organic matter; any thinner and the weeds will continue to come through. However, this should not be done before May so that the soil can warm up first. The third is to leave bare and hoe, but being careful not to hoe off the emerging new shoots that will be next year's fruiting canes. If you leave it and do nothing, you risk a tangle of weeds that will soon become severe competition for both canes and pickers alike.

In the longer term, it would be possible to keep the surface clean by hoeing, but this would be a major job not to be undertaken lightly, and not really in keeping with organic principles. Mulching the rows with organic material from May onwards would be better, but the growth of new canes makes its laying down quite a business. The mulching material can come from the vegetation between the rows, which can be a green manure of white, sweet clover (*Melilotus alba*) or subterranean clover (*Trifolium subterraneum*) sown at the rate of 30 and 40kg/ha respectively, and cut twice in the season.

The alternative is to integrate the raspberries with a poultry enterprise so that the vegetation is kept down and the fertility is kept cycling, with the added bonus of a measure of pest control. Two factors govern the effectiveness of this method. First, the timing of the poultry's access must be carefully controlled, to ensure that they neither harm the young shoots as they emerge nor eat the fruit as they ripen. Second, the stocking rate must be sufficient to control the vegetation properly without damaging the raspberries. This probably means either having other areas nearby where the birds can go when they are not in the soft fruit, or a timed succession of temporary batches such as fattening table birds. You would be able to reduce the initial input of manure by at least half with an integrated livestock system.

There are several training methods for raspberries. Perhaps the best is the one that uses two wires at either end of 90cm long crossbars, attached 120cm high to the upright posts. The new canes grow up inside the two wires, protected by, but not hindering the picking of, the current crop. They are then tied at an angle to either wire, about 10cm apart. A variant is to actually twine the canes around the wires rather than tying them. This makes the whole structure much stronger and more self-supporting. With narrow tractor access, rows need to be about 270cm apart. This system is not suitable for the less vigorous varieties like Malling Jewel, for which more conventional methods are needed. Autumn Bliss, on the other hand, has strong canes and needs very little staking.

The worst pests are raspberry beetles and aphids. The former can be controlled with two sprays of derris, about ten days after flowering starts and again ten days later, but mind the bees. Poultry will help too. Aphids are serious because of the viruses that they carry. A 5 per cent tar oil wash, thoroughly applied in January, should keep their overwintering population in check. Spur blight and cane spot are the two most prominent diseases, but should not cause serious problems unless the land is badly drained or too acid. Cut out badly diseased canes and spray the rest with Bordeaux mixture at bud-burst, and again ten days later.

Black Currants

Black currants are of relatively minor importance, with most of the conventional crop going for processing. There is a limited demand for fresh sales and for pick-your-own, but there is certainly an opening for a small organic processing industry to develop, the beginnings of which can be seen in the yoghurt business. To get a spread of cropping, grow Blackdown, Ben Lomond and Malling Jet. All three are mildew resistant and the last two are late flowering.

Land for black currants needs to be of good quality, deep and moisture retaining, but equally important is good shelter and relative freedom from late frosts. Ensure that perennial weeds are absent, as for raspberries, subsoil in dry conditions and spread a good 100t/ha of manure before planting in November or even October. Unless you are working on a large scale, you will not be machine picking, so spacing can be anything down to 1×2.5m. Alternatively, plant three rows 1.5m apart in beds

3m apart, and grub up the middle row after four crops. More continuous rows with 0.6m spacings between plants are possible, but might be more difficult to manage organically.

Every spring spread another 20t/ha of well rotted farmyard manure around the plants. An equivalent in organic fertiliser will just about do, as long as it is not low in potash, but black currants do like their organic matter. Seaweed foliar spray, as with raspberries, is also beneficial. The growth habit of black currants makes weed control as difficult as with raspberries, so the same techniques can be applied, the best of which is mulching. The poultry option is less applicable because the vegetation, being much lower, would be more vulnerable. Pruning is entirely conventional.

Big bud gall mite is the most serious pest because it carries and spreads the reversion virus. Diseased plants must be dug up and burnt. The mite can be controlled by spraying with lime sulphur or derris-pyrethrum mix just before and after flowering. For the smaller plantations, you can look for, remove and burn the big buds that are the mites' homes, at pruning time. Aphids may be a problem, and can best be prevented by a thorough winter tar oil wash in early February. Leaf spot is the worst disease and is symptomatic of too much nitrogen, among other things. If it is bad enough to defoliate the bushes, burn the dropped leaves (you might be able to do this *in situ* with a flame gun), spray the bare branches with Bordeaux mixture and miss out the next annual manuring.

Gooseberries

This is an even more minor crop, but there is a market for a small amount of fresh sales and for pick-your-own. Its low importance is reflected in the lack of new varieties available. Invicta is one and replaces Careless from which it was bred. For less yield but more taste, try Whitesmith or the red Lancashire Lad. All are mid-season and are resistant to mildew. The very new Greenfinch, again of Careless descent is also resistant to leaf spot, and might be worth considering.

In the past, gooseberries (and red currants, below) were sometimes grown under top fruit, particularly plums. This practice is now discontinued because of the specific spraying programmes that each require with conventional methods. However, it might be worth investigating such a combination for organic culture, where the management can be more versatile and the diversity is likely to be beneficial to both crops.

Treat the soil as for black currants, except with only about 50t/ha of well rotted farmyard manure spread in the autumn and with the addition of 500kg/ha of rock potash. Plant in October at 1×2.5m, but again suiting the spacing to your mechanical requirements. Annual manuring thereafter need be only 10t/ha, but with the same seaweed spraying as above, and repeating the potash dosage every four years. Keep an eye out for magnesium deficiency, to which gooseberries are rather sensitive, and use kieserite to correct it, or dolomite when liming.

Weed control is somewhat easier because gooseberries are grown on a 'leg' up to 20cm long. Mulching is again the best, although hoeing is more possible, and even a longer-term polythene mulch might be worth considering. Pruning is conventional, but delay it until February if bullfinches are a problem, to give the plants a little more cover. If mildew still strikes, despite the resistant varieties, use a regular spray of 0.25 per cent washing soda from March onwards. Other pests and diseases should be of little trouble to the organic grower, or can be controlled either as for the other soft fruit, or by standard organic means.

Red Currants

Of even more limited demand are red currants. Until an organic processing industry is established, the only likely outlet of any significance is the pick-your-own market, and even this is very small. For a good succession, try Jonkheer van Tets, Laxton's No. 1, Stanza and Redstart. You might also find a small space for some white currants, which are variants of the red. Go for White Grape or White Versailles. Their treatment is exactly the same.

Red currants can manage on slightly lighter soils, but still need it to be deep and moisture retaining, and in a sheltered position. After dealing with the perennial weeds and subsoiling, manure as for gooseberries and plant in November at about 1.2×2.5m, or more for machine access. Red currants are also grown on a leg, so weed control is easier than with black currants.

The only disease of note is coral spot which needs to be cut out and burnt. Otherwise, leaf spot and mildew can be treated as for the other soft fruit above. Similarly, a tar oil wash in February will keep aphids under control.

Blackberries and Loganberries

There is always talk of new cane fruit coming on to the scene as breeders play around with new parents and different crosses from this vigorous family. Somehow they never seem to gain the prominence that one might expect. An exception, though an old one, for it is over a century old, is the loganberry. Taming the blackberry has been another challenging breeding exercise. Both have a very limited market however, restricted mainly to pick-your-own sales until an organic processing industry gets going. The only other ones to merit a mention are the tayberry and its close cousin the tummelberry.

LY59, or the thornless L654 are the loganberries to go for, and Oregon Thornless for a blackberry. If you want a really vigorous windbreak as well, you could grow the thorned Ashton Cross. However, Oregon Thornless would also do. Both berries are considerably less demanding than raspberries and only need a moderate amount of manure, say 25t/ha and 500kg/ha of rock potash, incorporated with the pre-planting preparations.

Being so vigorous, these plants need strong supports on which to be trained and planting should be as much as 240cm apart for Oregon and 300cm for the loganberries and Aston Cross. The rows should be at least 180cm apart, or more for machine access. Weed control can be the same as for raspberries and, because of the spacing, should prove the easiest of all soft fruit, but include a mulch of manure every spring. Prune out all the old canes after they have fruited and burn them. In their place train the new growth, which should have been gathered loosely together out of the way as it has developed. Pests and diseases are basically the same as the other soft fruit, particularly raspberries. Treatment is therefore also the same.

Rhubarb

Here is a crop that many growers have in an odd corner, covered in weeds most of the year and generally neglected. Luckily it can take such poor treatment, although obviously it does not thrive on it. However, this does pose a dilemma to the serious rhubarb grower because everyone is out to sell their crop, including the gardeners who always have more than they need, and the market soon becomes glutted. Pick-your-own sales may be one outlet, but prepacking for supermarkets probably offers the best prospects, as this is entirely undeveloped at present. The early, forced crop is also wide open, but is most easily developed in association with a larger maincrop area. Timperley Early is still probably the best early variety and Victoria or The Sutton for maincrop.

Rhubarb will grow on almost any soil but likes good drainage and plenty of organic matter. Ensure that the ground is clear of perennial weeds, as with the soft fruit above, subsoil and incorporate a generous 75t/ha of farmyard manure (more on light land) with your final soil preparations. Then plant the sets, which are basically the divided bits of older crowns, in October for the earlies or after Christmas for the maincrop varieties, at 70×90cm. You can plant into ridges if your soil is particularly wet or if you are going for early production. Pick nothing in its first season and only lightly in its second, but hard thereafter. Every February, mulch the crowns with a good 50t/ha of farmyard manure. A generous tonic of dried blood, when you have finished cropping in June or July, will also be appreciated.

It is probably best to establish a rotation with the rhubarb so that one fifth or one sixth of the area is dug up each year for forcing and discarding, or dividing and replanting elsewhere. For the earliest forced rhubarb, lift the crowns in the late autumn and leave them upside down in the field to get frosted. Then, bring them into a tunnel in early January and settle them in very close together, just nestling in the soil which should be well moistened. Cover with black polythene, probably over cloche hoops to leave a growing space of 60cm. You should be cropping from the middle of February for about five weeks.

The next batch can be forced in the field. In the middle of January mulch well and cover the rows with black polythene cloches. This will advance them by about three weeks. Once the main bulk of the crop has started, these can be removed so that the forced crowns can recover without undue stress. A deep straw mulch will have a lesser effect but is an advantage in terms of the quality of sticks produced.

Irrigation is certainly an advantage, but is not entirely necessary. Weed control is necessary however, and needs careful attention despite the good covering that rhubarb produces. There are not really any serious pest or disease problems, but just about all rhubarb is infected with viruses to some degree or other. Therefore make sure that you maintain good vigour by choosing only the best crowns for dividing and replanting if you use your own stock.

Grapes

There are two organic vineyards in this country and many in Europe. They can and do produce wine of the highest quality. Organic viticulture is entirely possible, despite the supposed sensitivity of vines to pests and diseases. For someone who wishes to make this his particular speciality, who has land of the right quality (which means sunny, sheltered and free from late frosts), and is prepared to put considerable effort into the processing of his products and their marketing, there is hardly a more rewarding crop to embark upon.

The growing of grapes is a subject far beyond the scope of this book. However, it is another possible option for the small farmer who is lucky enough to be in the right situation. One other thing should be mentioned: some say that the British wine market is nearing saturation due to the high price that has to be charged as a result of a relatively marginal climate for grapes. The price certainly puts British wine in competition with far superior continental wines, so quality is paramount in this product.

Blueberries

For the really ambitious fruit grower who is looking for something new and completely different, then the blueberry must be the answer. On the conventional scene it is beginning to be heralded as one, if not the only one, of the new fruits of the future. Grown widely in America, it has a long productive life and from a culinary point of view is extremely versatile.

TOP FRUIT

If you are going to take the plunge into top fruit growing, consider the degree of intensity of the enterprise you want, and hence to what extent it can be incorporated into the rest of the farm. There are certainly arguments in favour of a more integrated approach in an organic system for a perennial crop that is so beset by pests and diseases. Other points to bear in mind are mainly economic, and the top fruit sector has been particularly pressurised in this direction, leading to the development of highly intensive orchards, totally dead and bare soil and greater reliance on chemicals and pesticides than almost any other sector of agriculture.

The range of intensity extends from pure stands of dwarf bushes, through stands of mixed varieties or even species, to half-standard or standard orchards, grazed by livestock and finally to field boundary plantings, or just trees, spaced throughout the holding. The main drawback of the less intensive systems, apart from the extra labour that they demand, is that they can easily be neglected as an enterprise, and the one thing that top fruit need is careful and expert management.

Apples

Apples are the most widely grown top fruit, and for very good reasons. They are less temperamental, hardier, more versatile and have a greater number of varieties to suit more conditions than any other tree fruit. They will fit any intensity of system and have varieties that will produce apples for eating and cooking for a good nine months of the year.

Choice abounds; your first choice will be of the system you wish to use. That will automatically lead you on to the second, which is the choice of rootstock, for that will influence the size that any plant will reach. Dwarf plants begin with M9 followed by M26, then MM106, which is a semi-dwarf, and MM104 which is for larger trees.

Next comes choice of varieties; a number of factors must be taken into account. One of the most important for organic growers is disease resistance, and unfortunately our most popular apple, the Cox's Orange Pippin does not rate very highly. However, there are any number of new and old varieties that are prolific, disease resistant and of good flavour. You must, however, investigate what varieties suit your own conditions. Remember also that the chosen trees must be compatible for pollination at the right time of flowering. A good succession, in reasonable growing conditions, might be provided by Discovery, Katia, Sunset or Egremont Russet, Chivers Delight or Holstein, Winston or Idared, Pixie or Malling Kent. Of the cookers, try Golden Noble and Newton Wonder.

Apples like their soil not too rich, but well drained and free from late frosts. Plant carefully, preferably in November, into a generous hole and incorporate compost or well-rotted manure with the soil (at a ratio of about 1 to 4) as you fill in, firming as you go. Sprinkle in a handful of rock potash too. Stake well, especially the larger trees. Spacing will depend entirely on the system, and will range from 1×2m (or more for machine access) for spindle bushes, to 5×5m for standards. Hoe and mulch, keeping a good metre of ground around the young trees clear of weeds.

Every spring a mulch of manure can be thinly spread around this clear area, but not too close to the trunk. The pathways between rows can be sown down to rough meadow grass, or similar fine grass, and wild white clover. After a few years, depending on the size and vigour of the plants, the whole orchard can be grassed over, but again not right up to the trunk. Keep tightly mown or grazed.

Of the long list of pests and diseases, codling moth can be controlled by tying bands of sacking around the trunks near the ground in June and removing and burning them in October. Grease bands, put on when the sacking comes off, to stay until June, will prevent several moths from climbing up to lay their eggs. Apples infested with apple sawfly fall off by June. These should be collected and destroyed, or fed to pigs. Hoeing round the trees in winter will also help to expose their hibernating caterpillars.

There are several beneficial insects that can be encouraged in the orchard, firstly by being very hesitant about spraying with any insecticides. Tar oil wash in winter is unfortunately to be included in this hesitation, as it will kill the predatory bug *Anthocoris nemorum*. Similarly, lime sulphur and other sprays around blossoming time in May will kill the black-kneed capsid, which is another good predator. If there are aphids, then try growing buckwheat or flowering umbelliferae around the trees to attract more hoverflies. One summer pruning in July will also help by reducing their new growth food supply.

Most of the varieties listed above are fairly or very resistant to scab. Further reduction can be ensured by removing the fallen leaves in December or chopping them up so fine that the worms can take them down before the spores are able to spread about. Inspect the trees after pruning to catch any diseased parts that have been missed. Do not winter prune until at least February; this will then miss the brown storage rot fungal spores that are no longer around.

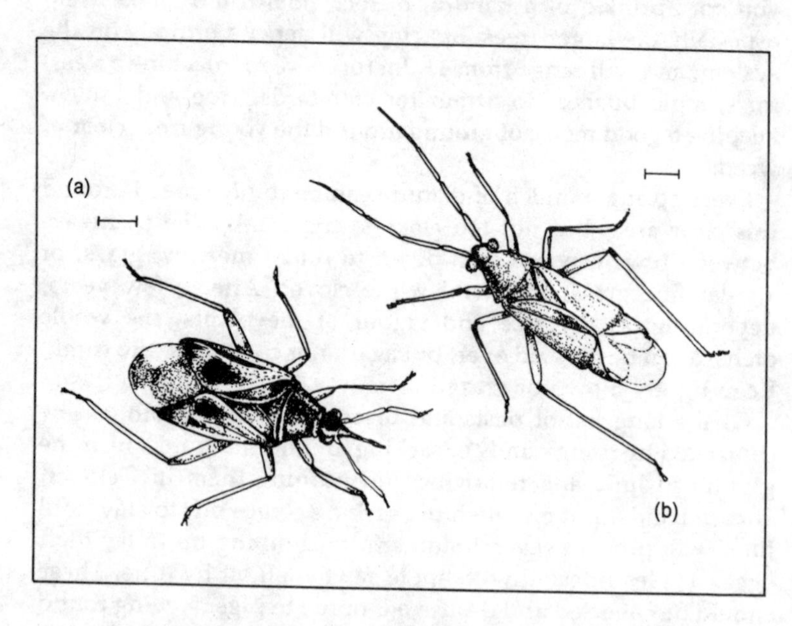

Beneficial insects in orchards: (a) Anthocoris nemorum*; (b) Black-kneed capsid. (Insects shown 9 times life-size.)*

Pears

Pears are not so easy to grow as apples. They flower earlier, so need a warmer and more sheltered position, and despite needing richer and generally better soil, they still yield lower. Rootstocks should be Quince C for the most dwarf, Quince A for semi-vigorous growth, and their own, or pear stock for larger trees. Try the varieties Jargonelle, Glow Red Williams, Nouveau Poitou, and Josephine de Malines for a succession of fairly scab-resistant trees, and Catillac for a cooker. Many varieties need at least one and sometimes two compatible pollinators. Therefore, when choosing varieties for your particular area, check carefully for this as well.

Much of the advice about apples can be directly applied to pears, including planting, pruning and control of some pests and diseases. However, pears prefer better soil and need better manuring during their life, so make the spring mulch of manure up to a third as much again. Also be much more reticent about grassing down under the trees, as this tends to provide more competition than they can take. Picking is more critical too. The early varieties are better picked on the early side and the later ones are better picked as late as possible.

Scab is one of the worst diseases, but resistant varieties, careful pruning, dealing with fallen leaves, and Burgundy mixture if needed as a last resort should be sufficient. Good housekeeping, by removing and burning diseased parts, must be employed against leaf blister mite and pear midge if both are to be tackled early enough; leaf blister will also need Burgundy mixture. Leaf curling midge will go with summer pruning, as will aphids. Pear thrips can be controlled with pyrethrum in the evening (to avoid the bees). Lastly, fireblight has no organic answer, but it is a notifiable disease and must be reported to your local MAFF offices.

Plums

Most organic plums are the surpluses from peoples' back gardens, so the potential for commercial production of these relatively exotic but also prolific fruits is great. However, they flower early like pears and therefore do best in areas that do not get late frosts. They need plenty of light and warmth for the best flavour, and prefer good, rich, moisture-retaining soils with lots of organic matter. They are probably more suited to the eastern counties of England rather than the wetter west. Plums also prefer to be larger trees with plenty of space, rather than restricted growth.

St. Julian A is a semi-dwarfing rootstock suitable up to half-standard size and Brompton or Myrobolan B are vigorous rootstocks for standards. Good varieties include Laxton Early Gage, Victoria, Edwards or Jefferson, and Anna Spath or Ariel for a reasonable succession. Good cookers include, in order of ripening, Rivers' Early Prolific, Czar, Pershore Egg, Warwickshire Drooper and Marjorie's Seedling. Once again be aware of the need for compatible pollinators.

Manuring can be the same as for pears, with the addition of 750kg/ha of calcified seaweed to the orchard each year. The ground should preferably not be grassed down around the trees, but they do benefit greatly from generous mulching. Pruning is different with stone fruit. It is really just a question of taking out crowded and weak growth and any shaping that might be needed. The most important thing about pruning is to leave it until May, when the sap has started to rise again and the danger of silver leaf getting in to the wounds is lessened. Mild summer pruning at the end of June is possible if you need to curb the vigour of a particular tree.

If silver leaf has caught hold, cut back to clean wood and if severe, slit the bark with a sharp knife from the cut right down the branch and trunk to ground level. Cut back anything infected with powdery mildew, bacterial canker or other dieback symptoms and spray with Burgundy mixture in January. Any cutting done outside of the safe time from May to 15th July needs sealing with a good protective paint. Aphids can be controlled with a winter tar oil wash in January and pyrethrum in the summer.

9
Livestock

It may seem strange that a book on organic farming which advocates a balanced approach only has one chapter on livestock to four on crops. This is not to denigrate the importance of livestock, nor to underestimate the differences between types; rather it is because, firstly their role in a small farm is likely to be subsidiary to the more profitable crops, and secondly organic livestock husbandry is less different from the conventional than is the case with crop husbandry.

The criteria for organic livestock husbandry are nevertheless very stiff, and there are in fact very few Symbol-holding livestock farmers. Much of this is to do with the strict rules about organic feeding, which many farmers, and particularly small farmers, have great difficulty in meeting. There is therefore less of critical organic importance to say about livestock than there is about crops.

So much of organic livestock farming is common sense, humanity and good husbandry; unfortunately so much of conventional livestock treatment goes totally against these three things. Common sense, humanity and good husbandry should be regarded as the base line from which organic livestock production starts.

The ideal organic farm should maintain an equilibrium between livestock and crops. However, on smaller farms economic pressure will tend to tip that balance in the direction of crops, as these are likely to be more profitable. This is made easier if there is access to outside sources of manure to provide the nutrient cycling and fertility that a farm's own livestock would otherwise contribute. Nevertheless, livestock on a farm do complement the arable side and make the smooth running of the whole that much easier and more complete.

It is all too easy to allow livestock enterprises that are secondary to arable ones to just trickle along, supporting the rest of the farm, but only breaking even (or worse) themselves. A badly run sheep flock with low lambing percentages, a couple of sows in some old buildings round the back or a few old hens in the orchard do no justice to the farm, the farmer or indeed to the animals themselves. Livestock demand greater and more constant care and attention than crops if they are to give of their best, and therefore it is the responsibility of all farmers who keep livestock, and particularly of organic farmers, to provide them with that. This is what good husbandry is all about. It is only sensible to build on this care and produce from it a profitable enterprise, rather than something that does not pull its weight. In the small farm situation there is no room for such slack.

STANDARDS

It is enshrined in the Soil Association Symbol Standards that the conditions under which livestock are kept should be suited to their physiological and ethological needs. This means that housing should allow free movement and have a maximum amount of fresh air and daylight, bedding should be adequate, and there should be regular access to pasture in fair weather. Breeding stock should not be permanently housed, and sow crates and other restrictions are not allowed. De-beaking, de-tailing, hormone implants or injections are all prohibited.

Organic Standards require that all forage produced on the farm must be organic, but that up to 20 per cent of the daily intake of feed on a dry matter basis can be of non-organic origin. These figures are currently being revised to accommodate the inclusion of some conversion grade feed. The non-organic portion should not contain synthetic amino acids, emulsifiers, colourants, antioxidants, urea, animal manures or antibiotics and other growth promoters. Seaweed powder, rock salt and natural sources of vitamins, such as cod liver oil and yeast, are the preferred supplements if any are needed.

Perhaps the most difficult area of organic livestock production is in the treatment of disease. Livestock tend to be more susceptible to disease than crops, mainly because they can be abused more, by an inadequately designed rotation, poor housing, overstocking and by being pushed too far. Intensive livestock production covers up such potential problems by recourse to the vet and the often routine use of anthelmintics, antibiotics and a host of other drugs. All these things should only be used in cases of emergency on the organic farm. They should certainly not be routine, but they should not need to be, for this would indicate a system that was out of balance. The organic farmer must look at his system, to identify the causes of the imbalance and to remove them, both in the short and the long term.

As with crop production, the organic approach to animal husbandry requires a greater understanding of the situation in order to proceed and succeed. This is also true of the two forms of medicine that are recommended for use: homoeopathy and herbal medicine. These are very much in keeping with the organic philosophy because, in common with all 'alternative' medicines, they regard and treat the organism as a whole. In the right hands with correct and early diagnosis (and this may rely on different factors from those required by conventional allopathic treatment) they are safe, cheap and very effective. You would do well to familiarise yourself with either or both as much as possible.

Briefly, herbalism uses herbs that are specific to a particular condition, in various forms from concentrated tinctures to fresh from the plant, to help the body's own defence and healing mechanisms operate at their optimum. Unfortunately, these various forms are not always easily taken by livestock, which does sometimes make herbalism difficult to put into practice. Including herbs in your pastures as preventive medicines is valuable, however, not only as a source of natural minerals but also for the tonic and cleansing properties that they have. Garlic (the wild variety) and mustard are classic examples and should be available to stock for this reason, although discretion must obviously be used with milking animals and garlic. On this level the difference between herbalism and diet is virtually non-existent; a good and correct diet is one of the first steps to healthy animals.

Homoeopathy is much more difficult to explain. Its principle is rather bluntly summed up by the phrase 'the hair of the dog that bit you'. A substance that induces a reaction in a body has, at infinitesimal dilutions, the power to heal that same reaction. Furthermore, a substance that produces a similar reaction also has that power, and the greater the dilution, even to the extent when there is none of the active ingredient left, the deeper is the level at which it works. The dilution, called 'potency' in homoeopathy, is achieved by a critical and important process, and is denoted by a number after the name of the remedy. The higher the number, including the Roman numeral, the greater the dilution and hence power. The most common numbers to look out for are 6X, 30C, 200C and 1M.

It is important to fit the right potency to the particular ailment, or it might otherwise not be enough or indeed can even produce an over-reaction. Generally speaking the higher the potency, the fewer tablets that need to be administered, so a 6X potency might be given one an hour for a week, whereas 200C might only be given two or three times a week for perhaps three weeks. It all depends on the acuteness of the symptoms and how deep-seated they are. Treatment of different types of animals is exactly the same for the same symptoms, except that frequency of dose is increased as the size of the animal decreases.

FOOD

The ruminant stomach is designed to eat grass and digest fibre. The more you fill it with concentrates, grain and high-protein feeds, the further from its natural diet you are taking it, and hence the more problems you are likely to encounter. This is most clearly demonstrated with dairy cows, as these are probably pushed the hardest and for longer than all other ruminants. The rumen bacteria are altered and this lowers rumen pH and produces damaging metabolites which can escape through the rumen epithelium. Several seemingly unconnected areas of the body are then affected. The already stressed liver becomes more damaged and over-fat, impairing fertility. White blood cells that are part of the body's defence systems, are prevented from working both by acidosis and by free fatty acids in the blood; this lowers resistance to infectious diseases including mastitis. Histamine and lactic acid reach the feet and cause the foot diseases that are so common now. It is not surprising that the average life expectancy of a dairy cow in a herd is now less than four lactations, whereas in organic herds it is at least double that.

So the first principle of feeding ruminants is to feed them with a maximum amount of fibrous foods, only topping up with concentrates where needed. You should try to produce all the fibrous foods on the farm, for it may be very difficult to obtain organic forage in the locality. Concentrates are another story. It is not easy to find organic grain for livestock, as they have to compete with the vast human demand, but I am assuming that the small farm is not large enough to provide for its own concentrate needs and will buy in from other, preferably organic, sources.

The situation with non-ruminants is totally different. Fibrous food for them is more of a luxury than a basic dietary component. It is beneficial for the complete diet, not only for its nutritional contribution but also as an aid to healthy digestion, rather like roughage for humans. What is equally important is the exercise and fresh air that the animals enjoy by foraging outside.

Pasture

It is likely that there are three main types of pasture on a farm. The first is permanent pasture that has been down for more than about eight years. This may be because it is on land that is too steep or too wet to cultivate, because the soil cannot take anything but permanent cover, or because of the previous occupation of the farm. Usually permanent pasture has a rich diversity of plants, which makes for good grazing quality, but it may also be very run down and low yielding.

Improvements can be made just by management, and indeed the continued quality of permanent pasture depends entirely on management. Make sure that the grass does not get overgrazed in summer by overstocking, nor poached in winter by having stock on too long or too late. Ensure that nutrient levels, particularly phosphorus, are sufficient, and pH is about 6.5. Spread any manure or organic fertiliser in the early spring to avoid deterring the clover. Alternate cutting and grazing, and have a regular system of topping.

Improvement by reseeding will be quicker and it need not entail ploughing up and starting again from scratch. Graze down very tight and harrow heavily until there is some bare soil on to which the reseeding mixture can be sown. Harrow some more to incorporate the seed, roll and leave until the new plants have established. A better alternative is with strip seeding machines that are available on contract. The reseeded strips colonise the rest of the pasture once they themselves have established.

Leys of medium- and long-term duration, including the conversion ley, are usually part of longer arable rotations and constitute the second type of pasture. The length will depend on the type and state of the soil as much as on plans for the rest of the rotation. Sowing is best around April or August, although there is considerable leeway on either side if conditions suit. Clover seed, in particular, is very small, so try to take a weed strike before sowing and ensure that the seedbed is fine and firm. For an average, suitable mixture *see* Chapter 3.

A catch crop in the year of establishment can often be taken by either undersowing the grass mixture, preferably in a spring cereal crop, or sowing it with about 4kg/ha of Italian ryegrass to give early grazing, or even with forage peas. Do not let the nurse crop get out of hand or it will smother the developing ley. Once established, management is all-important to ensure that the sward stays productive with the right balance of herbage.

The third type of pasture is the short-term ley. This is usually just a one or two year break in the middle of a crop rotation. The grass and clover need to establish quickly and grow fast. The balance of species and their management are not so important, and so the mixture is usually just red clover and Italian ryegrass, used principally for conservation.

All pasture will benefit from manure, if you have it spare, at the rate of 25t/ha in the spring, but unless the soil is in need of feeding it is usually sufficient to rely on the grazing animals and clover to provide the nutrient cycling and leave the manure for crops of higher value. Calcified seaweed at about 500kg/ha every three years will also be of benefit, particularly in wetter areas where the pH tends not to be stable. Rock phosphate might also be needed regularly in the wetter west of Britain.

Conservation Crops

Whether you opt for hay or silage, either clamped or in big bales, probably your best crop is the pasture that is not used for grazing. This gives greater flexibility for rotating the stock on your limited area and for responding to the effects of the weather on your production. Perhaps the only exception is lucerne, which can be regarded as a longer-term version of the Italian ryegrass or red clover break crop for drier parts of the country. It takes at least two years to come to full productivity, so it should really stay down for at least five years if its potential is to be realised. Start it off much like a ley and be very light on it for the first eighteen months. Once established it will benefit from heavy harrowing in the winter. Try Euver or Vela.

Maize for silage is a very greedy crop that does not suit the limited nutrient situation of a small organic farm. Cereal mixtures are possible, particularly if they contain legumes of some sort to contribute to the nutrient status, but perhaps should only be regarded as useful for special situations, such as when doubling as a green manure crop or as a nurse crop for a new ley.

Fodder Crops

Fodder crops for use in the winter and early spring are very beneficial to all animals, giving them something fresh at this inhospitable time of the year. The green crops, such as kale and other brassicae, and some cereals, generally have reasonably high protein contents and are good sources of β-carotene, both desirable in the winter. The succulent crops, such as turnips, swedes, fodder beet and other roots are good carbohydrate sources and generally have mild laxative properties that complement the rather dry winter diets.

Marrowstem, Thousand-headed and then Hungry Gap kale will provide a succession of fodder from early autumn until April. Sow in May after some 50t/ha of manure has been spread and a weed strike has been taken. Try the varieties Condor and Bittern for autumn and winter. Cabbage, rape and kohl rabi can also be used as forage, but are less common. Remember that all these are brassicae when it comes to working out the rotation.

Getting growth of pastures early in the spring is one of the main problems for organic stock farmers. One way round this is to have green crops specially grown for this period, and forage rye must be about the best. Sown in September or early October at the rate of 175kg/ha it will be ready for grazing in March, thus providing an excellent winter green manure and catch crop. Try the variety Humbolt.

Of the root crops, stubble turnips make an ideal catch crop for sowing any time during the spring and summer, remembering that it is a brassica. They have a rather low dry matter but a high crude protein, and will feed through until Christmas, being a useful complement to pasture at its shorter times. Try Barkant or Marco. Swedes need longer to grow but are hardier and have a higher dry matter, and are obviously flexible in their use – we eat them too. Marian and Angela are probably the best varieties.

Fodder beet is very closely related to the mangel, the traditional fodder root crop, but has now overtaken it and stands alone in its quality and yield. It can produce the highest dry matter, protein and energy of any fodder crop. Sow in April or early May, after a weed strike and an application of about 50t/ha of manure. Agricultural salt may also be applied at least a month before at about 300kg/ha. Lift in October or early November and store or clamp for use throughout the winter and spring. Try Kyros or Hugin. Mangels must not be used until the new year, but are otherwise grown the same.

CATTLE

Cattle are usually the most important type of livestock on an organic farm. This is unlikely to be the case on smaller farms, as their size makes them less adaptable to the scale of operations. Dairying is probably out, leaving the suckler unit, beef production and calf rearing as possible options. The first two are long-term ventures, with a return on capital taking up to three years. Furthermore, unless you sell the end product direct, there is not very much money in it, particularly compared to other enterprises on a relatively intensive holding.

Calf rearing can provide a useful and regular income if,

perhaps, a neighbouring dairy farmer wants his replacements reared and some contractual arrangement is devised between you. Any other sort of calf rearing is full of risks, especially if the calves are bought in the market from unknown sources. Veal production is a possibility, if it does not go against your principles, but it will not be organic as the special milk substitute that is used will constitute too much of the ration.

Suckler cows can be of any breed and are often dairy crosses for extra milkiness. The least intensive and most humane system involves leaving the calf on the cow for as long as possible, allowing natural weaning, but still keeping them together in 'family' units until they are finished. Calving is best just before the grazing season, so the calves are able to utilise the spring flush of milk. This may seem to be an unrealistic method, particularly on a small farm, but nevertheless it is profitably practised under organic conditions and is certainly the nearest to the organic ideal that any form of beef rearing is likely to get. The quality of beef will be the highest too, which is a very important factor when aiming for this quality market.

Multiple suckling is more intensive, and involves the risk of having to buy in calves. These, together with the true calf, can either be left on the cow for the duration of her lactation, or a shift system can be developed so that after three months another batch and then another can be reared on her. She should be able to handle four calves in the first batch, three in the second and two in the last. This will obviously involve more concentrate feeding than the single suckling system and probably a slower growth rate.

Straight beef production is usually either from weaned calves or from store cattle. Both have the risk of buying in stock unless you can find a regular and known source of animals, and the latter are particularly susceptible to fluctuations in price that can sometimes mean selling the finished animals for less than their buying price six months before. Nor can you really say that such beef is organic.

If you have to buy in calves, it is always a good precaution to administer homoeopathic nosodes for some of the common ailments, such as Salmonella, E. coli and virus pneumonia. (A nosode is prepared from the disease organism or diseased tissue, rather than from a substance that produces a similar reaction). Calves are the most susceptible to worms, and careful rotation of their grazing, exposing them only gradually to an increasing worm burden will help them build up strong immunity. Garlic and the root of the male fern (administered as a herb or homoeopathically) are good worm remedies, along with mustard, wormwood and many others. The chemical wormers that are allowed in specific cases are diethylcarbamazine and thiabendazole, and otyclozamide for liver fluke only.

New Forest Eye is becoming a problematic disease in summer, but there is a preventive nosode if it is severe in your area, and Kali Iodide 200 or Silica 200 respectively can be used for increasingly severe symptoms. Bloat is often feared on the lush clover sward that organic farmers rejoice in. The organic pastures seldom actually cause it, but linseed oil drenches can be a useful standby, and preventive too if given in advance of suspected trouble. Carbo Veg 200, followed by Lycopodium 200, will also treat it. Ringworm can be controlled by Bacillinum 200. Remember that hormone implant growth promoters are absolutely not allowed.

SHEEP

Sheep are the great complement to cattle. They do not share the same parasites, they graze in a totally different way and at the same time are great improvers of pasture. However, it is always said that a sheep's worst enemy is another sheep. They are very susceptible to worms and to foot problems, and are, in fact, not really suited to the relatively intensive conditions that we tend to impose on them. The golden rule is not to overstock sheep. Something in the region of seven sheep to the hectare should keep them out of trouble. It is also worth bearing in mind that for each cow you have room for, you can keep a sheep without further grazing.

As with cattle, it is best to avoid buying in stock on a regular basis, to lessen the risk of importing disease. This means that a permanent breeding flock is much preferred. Fat lamb production is the most common type of sheep enterprise, but sheep's milk is becoming increasingly popular and could easily become the major enterprise on a small farm if the horticultural side was not to your liking.

You might also consider breeding pedigree stock or producing rams, but this tends to be a much more skilled and time-consuming occupation which requires getting your name and your flock known about the shows. You will not get far if you simply regard it as a perk, for it has to be undertaken seriously. Wool is really only a by-product in this country unless you choose breeds like the Jacob or possibly the Black Welsh Mountain. Their fleeces are in good demand by home spinners and can therefore make considerably more than a marginal contribution to the returns of the flock.

The breed you choose will very much depend on what sort of land you have. The hills of Scotland should not be expected to support the downland breeds and vice versa. Whatever the system, you will need sheep that are hardy, sensible, prolific and good mothers; in short, ones that will look after themselves as much as possible, for even the best need good and constant management – they can be extremely demanding. There are many half-breeds and hybrids that are becoming increasingly popular these days, and a few imports from Europe.

Several breeds might be particularly suitable for an organic system: for the hill areas, try the Welsh Mountain, the Masham or Mule (the latter two being half-breeds). Lower down, try the Clun Forest and in the lowlands try the Southdown, the European Texel or the almost year-round breeding Dorset Horn/Polled Dorset. A possibility for crossing with perhaps the Southdown or Dorset is the Cambridge, a new cross-breed with particularly good prolificacy. The British Friesland, of Dutch origin, is probably the best for milking in our climate, although Polled Dorsets and others are often found in milking flocks too.

Successful management of the sheep flock depends on daily inspections for signs of developing problems. The first sign of an ill sheep is very often when it is dead; it can go down faster than any other farm animal. Therefore speed of detection is crucial and is not helped by the wool that can so easily cover a multitude of sins. Ensure that the ewe is in good condition or on a rising plain of nutrition as she goes to the ram, that the ram is of good quality and also in fine form, including his feet, and that he has to deal with no more than fifty ewes.

Keep a level condition for the first half of pregnancy and start to step up the feed gradually about two months before lambing, during which time much of the foetal development takes place. This, and gentle handling, is the best way to avoid stress as lambing looms closer, otherwise toxaemia and other disorders might start appearing. Phosphorus 30 should help with twin lamb disease if caught early, and raspberry leaves are excellent for facilitating lambing. Kale is one of the best feeds at this time.

It is becoming increasingly common to winter and lamb sheep inside these days. This is fine if the move is done well in advance of lambing, and that facilities, including sufficient space and bedding, are adequate. Winter shearing may also accompany this practice, and certainly reduces the stress on the animals. Once lambed, roots are good for building up the milk supply. When the ewes and lambs go out, divide their pasture up into six separate paddocks, allowing grazing for one week in each, with the lambs having access to the next one ahead. That will provide the maximum growth on the minimum area, with the maximum health.

At present, organic standards only permit two classes of organic lambs: those dipped with the new pyrethroid dip Bayticol and those sold before the rest of the flock are dipped with the conventional organo-phosphorus dip. However, the organo-phosphorus dip cannot be used on the ewes for the second dipping period. This situation is likely to change year by year, so do check current regulations. A major problem that presents itself with the pyrethroid dip is fly strike, for which it does not give long-term control. Derris or eucalyptus spray on a fairly regular basis should help, if your management can accommodate it.

The milking flock requires similar management, taking into account the higher nutritional needs of constant milking. However, the sheep's lactation is shorter than the cow's and the yield, although of considerably higher quality, is only about 16l per week. Perhaps the most important aspect of milking sheep is what you do with the milk. You will almost certainly have to embark on the intricate process of cheese and/or yoghurt making, and you will probably have to do the marketing yourself. Both will require time and careful attention.

Foot-rot can be a curse, but there is really no excuse for it as the bacteria that causes it cannot live in the pasture, off the foot, for more than two weeks. There is a homoeopathic vaccine, but initial treatment involving weekly foot-baths of zinc sulphate, foot trimming and rotation of pastures will cure the problem permanently. Thereafter normal routine foot trimming, to keep them in shape, and careful preventive treatment of all incoming stock is all that is needed.

Apart from the worming suggestions made for cattle, the best preventive treatment is to ensure that the sheep do not graze on the same pasture two years running, and that rotation of grazing and alternating with other livestock are organised as much as possible. The younger the stock, the more important this is. Orf can be treated with Acid Nit 200, and grass staggers with Calc Phos 1M and Mag Phos 1M together, however this is a less likely complaint on organic pastures. A lesser preventive dose before they go out might be better if any trouble is suspected. The normal five-in-one clostridial vaccine is allowed at present, but an alternative would be to take the vaccine, potentise it and then use it.

GOATS

Goats were the potentially great new small-farm animal of the 1970s just as milking sheep are in the 1980s. There are several fairly large commercial herds and hundreds of backyard goatkeepers now, so if you want to penetrate the market you will have more of a job to do than in the past. Nevertheless, there is still great potential if you are really serious and go for quality; the supermarkets are all waiting and the number of truly first-class goats' cheeses that are available is actually very limited. In other words, as with sheep's milk, so with goats' milk – you must get into processing.

There are four main breeds of goats and all have their advantages. The Saanan is the heaviest yielder, the Toggenburg is the best grazer, the Alpine is the best browser and the Anglo-Nubian gives the creamiest milk. There are British versions of the first three which are probably worth going for, unless you are a pedigree enthusiast, and they are closely related to their pure versions. Most herds are mixtures of all breeds.

As a contributor to the overall balance of the farm, goats are considerably less important than cows and sheep, partly because of the way in which they graze. In short, they would rather not graze, preferring to browse instead. They do not, therefore, improve pasture and are positively dangerous to hedges. The best environment is rough terrain, full of brambles, bracken and scrub. This is their home territory and they will systematically clear it in time.

Good quality grazing must also be provided however, if they are to yield up to their considerable potential. In fact this is something of a dilemma, for the high yields that goats are capable of (up to 9l per day) are only attained with high-grade feeding. Their stomach capacity is twice that of sheep, which rather points to the low-grade feed that they are designed for, and this can be exploited to the full with unnaturally high-quality forage and concentrates.

One aspect of the goat's natural diet must not be overlooked in this drive for yield. The mineral level in leaves, herbs and twigs is very high and goats need four times more of some minerals than sheep – phosphorus, sodium, cobalt, and iodine being the most important. If they do not have access to such mineral-rich forage, they must be provided with some alternative to make up the shortfall. Mineral licks are not really designed for goats, but seaweed meal, with the addition of steamed bone flour if the diet is calcium rich, would be ideal. Always have rock salt available as a lick.

Goat pasture should be abundant in herbs including aromatic ones. The normal ley mixture can have these extra herbs added and the existing ones can be doubled in quantity. Nettles make an excellent addition too, but are best grown elsewhere for cutting, as they are only readily grazed after they have flowered. Oats and vetches or lucerne both make ideal silage for goats, kale and cabbages are good succulent fodder and barley and oats are the best cereal feeds.

The three most important aspects of goat husbandry, besides feeding, are housing, fencing and worms. Goats are sensitive creatures and hate rain, so they must always have access to cover. They also like to sleep off the ground, and should be provided with a platform some 30 to 40cm up for that purpose.

Draughts need to be minimised, especially down draughts. Good, strong fencing is a necessity because goats are both agile and inquisitive, and they will soon find a weak spot to squeeze through if given half a chance, usually to the detriment of your garden! Electric fencing is ideal, preferably three strands on top of half-height wire sheep netting and powered by a mains unit.

The main ailment with goats is worms, and an infestation very soon shows itself by producing taints in the milk, along with the more normal signs of unthriftiness. Goats are as susceptible to worms as sheep, and they share the same ones; avoidance and treatment are therefore the same as for sheep. Mastitis is always a potential hazard for milking animals, much reduced when they are hand milked and when their yields are not forced too much. Goats will usually dry themselves off about six or eight weeks before kidding, so reducing the risk then. If problems arise after kidding, however, with a hot, throbbing udder, try Aconite 1M and then Belladonna 1M. Use Bryonia 1M for a hard udder and Calc Fluor 30C for any small lumps that remain afterwards. Kidding itself usually proceeds without trouble, certainly much better than with sheep. Immediately after the birth, give the nanny a good handful of ivy to help her cleanse.

Goats are prolific breeders, often producing three or more kids. This is somewhat offset by the long lactation, which can continue for two or more years if given the chance. Many goatkeepers therefore kid their goats every other year, which helps to even out the overall herd yield. This practice may start to become less popular as the market for kid meat increases. There is not yet a definite organic market, but reasonable prices are paid, which makes the process of kidding and raising the billy kids for meat a possibility as a sideline. The billies must be castrated if born late however, because they are early developers.

PIGS

There is no doubt about the place of pigs in an organic farm, although it may be totally different from the role they play on most conventional farms, where they are confined to factory units of cramped, insulated boxes that produce meat (of sorts), disease and volumes of almost toxic slurry. Pigs are wonderful converters of waste products into meat and manure. They are also wonderful converters of waste ground as the first step to reclamation. They will graze pasture but they prefer to dig it up and they will reduce any crop residues to manure in no time at all. Left to their own devices, they are clean, friendly, intelligent and healthy. It is the sordid conditions to which they are normally subjected that earn them their undeserved reputation.

One of the problems about pig farming is the 'pig cycle'. Pigs can be bought and bred up into breeding units in a very short space of time. Similarly, it is easy to go out of pigs. The supply is therefore always fluctuating in response to the price, but always a little late. As it drops because of excessive supply, people go out of pigs; then there are not enough, so the price rises again and everyone goes back into them, causing a further glut and consequent fall. Despite this being common knowledge, everyone falls for it.

You have two options to get round the problem. First, once in pigs, stay with them through thick and thin, for the situation will right itself in due course. Find good markets that require consistency and quality, proper curing for instance, or establish a sound local trade that will stand up to the gluts better than the fickle local wholesale markets. Second, go into pigs when the price is at rock bottom and you can buy in cheaply and get out when it is at its zenith, selling out well. Obviously, from the point of view of maintaining a balanced farm, the first alternative is the preferred one.

Now you have to choose whether you go for a breeding unit, a fattening unit or both, and if fattening, whether pork or bacon. The breeding unit requires greater management skill. It is probably less affected by price fluctuations but does rely on having a good arrangement with a buyer to take on the weaners. Fattening is the more ruthless business, where insulated sweat boxes churn out cheap meat by the processed ton. You will never be able to compete with that, so you must find a market that will accept the free-range or non-forced animal and will be prepared to pay a realistic price for it. In the unlikely event of you being able to grow or purchase sufficient organic feed for them, you will find that the organic market is wide open.

The difference between pork and bacon is one of weight, breed and whether you want to sell whole or process as well, for extra value. There are also cutters, in between the two, and heavy hogs at the top end which mainly go for manufacturing. As for breeds, those suited to more intensive conditions are best avoided, as they will not be hardy or sensible enough to thrive under the harsher conditions. The Welsh is a good compromise between modern efficiency and hardiness. The British Saddleback is an ideal outdoor pig, but because of its colouring, is best for crossing. A hybrid of the two makes a very useful animal, either to finish or as your basic breeding sow. Of the old breeds, the Gloucester Old Spot is prolific and able to live on very little, earning it a reputation as the orchard pig, and the Tamworth is the best digger and the tastiest pork, but has small litters and a slow growth rate.

Probably the best system that combines health, humanity and efficiency is to have the sows outside on free range, farrowing in spring and autumn to avoid the worst of the

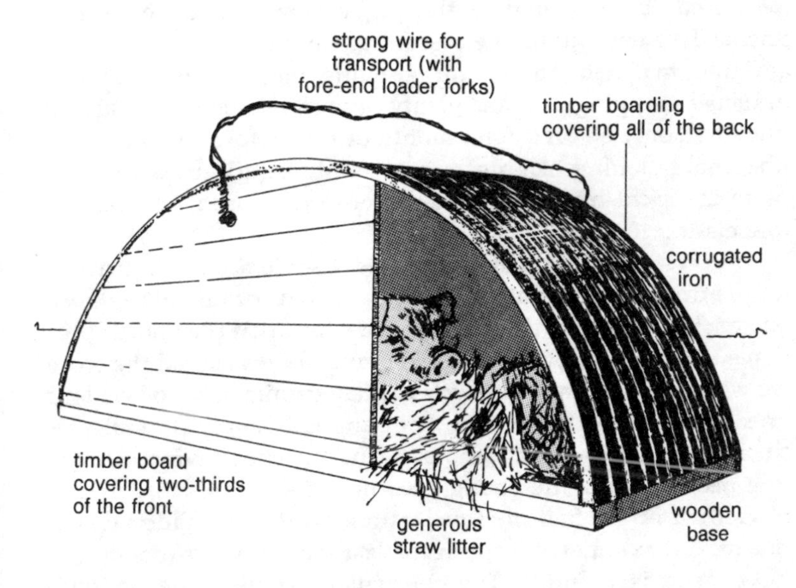

Outdoor pig ark.

winter weather, weaning at the traditional time of eight weeks, growing on the weaners outside in summer, or inside in winter, and finally fattening them off inside.

Six sows will keep a boar, but twenty is the maximum that he can handle, and the stocking rate will be about nine or ten sows to the hectare. Depending on whether you want them to graze or to break up the pasture, they will be ringed or not. Remember that they are excellent pioneering animals, so their best place might be on leys in their last year, where they will do the initial cultivations. Beware of the damage that they might do to the soil structure in winter. For this reason the lighter and freer-draining soils are more suitable. In any event, move the feeding troughs every other day to avoid poaching. Arks are perfectly adequate for housing and, with plenty of straw litter, for farrowing too. The more you rely on the sow's own instincts, the better they will serve her.

Piglets reared like this will have no need for the standard iron injections, or other props of the intensive system. Start creep feeding at the normal time of three weeks. Very early weaning is not in keeping with organic practice. Eight weeks is therefore preferred, by which time the piglets are able to look after themselves without undue cosseting. After that, the piglets can go into matched groups for growing on. Ground that has finished cropping and has plenty of crop residues is ideal, for they will only be on it for a month or two before coming in for the final fattening. All pigs can be easily kept in by an electric wire 25–35cm off the ground. Arks can be used for housing piglets too.

This system loses out to the intensive systems only in that the pigs are not in insulated housing. Unlike ruminants, whose stomach activity provides all the body heat that they need, pigs have to expend a lot of energy keeping warm, and all the more so with the leaner and less well-insulated animals of today. This means they need more food. Foraging will amply provide this in summer, but will be completely inadequate in winter, hence the need to bring the young ones inside at this time.

Getting on for half the sow's rations will be provided by the pasture in summer, except in the last month of pregnancy and when she is in milk. The shortfall must be made up with provided feed, consisting of cereals and a protein supplement.

Barley is the best grain and whey, skimmed milk or meat and bone meal are the best protein sources. This last source can be fed freely, as sows will not overfeed on it. The shortfall will range from less than 1kg when dry, to about 3kg when in full milk. One other feed worth mentioning is acorns. If you are lucky enough to have an oak wood, turn pigs out in it when the acorn crop has fallen, for it is one of their favourite foods and they will thrive on it, with perhaps the addition of some cereals.

The creep feed needs to be the highest in protein of all the feeds that you give, up to 20 per cent. For the small amount that they eat at this stage, it may be worthwhile buying in a proprietary concentrate, remembering that this will contain copper. Otherwise, the grower mix can be supplemented with 50 per cent more protein supplements, together with maize, cod liver oil and skimmed milk powder. Obviously it must all be of the highest quality.

The grower ration should be about 17 per cent protein. Feed it freely until the pigs reach 45kg in weight, and thereafter twice daily, as much as they will clear up in fifteen minutes. Bean meal, soya or milk by-products, at 16–20 per cent of the ration, will provide the main protein. This proportion means it can be from non-organic sources. Barley meal can make up the rest of the feed, perhaps with some other cereals. Alternatively, substitute the cereal portion with any of the roots (but in greater quantities). Fodder beet (×5), swedes (×8) and potatoes (×4 if cooked, more if raw) are probably the best. Finally, a finishing ration can be given from about 60kg if the pigs are going on to bacon weight. This need be no more than 13 per cent protein, which means only a minimal amount of protein supplement. It goes without saying that all animals must have access to plenty of clean water, and in hot weather pigs love to cool down with a good wallow.

Pests and diseases should not really be a problem to healthy outdoor pigs. They are naturally hardy animals who only succumb when confined in the unnatural close quarters of intensive units. Worms and skin parasites are likely to be the only problems you will meet outdoors. The former can be controlled by good rotations, the normal herbal remedies and green potatoes may be effective too. The latter are killed by derris, and once the herd is clear, they should not reappear.

POULTRY

Poultry are an ideal form of livestock for the small farm. From purchase to production is a short enough period of time not to strain the bank balance, the unit of production is small (that is, the birds themselves) and hence versatile, and there are any number of possible systems to suit almost any situation. The main bird is obviously the chicken, for both eggs and the table, but do not forget geese, ducks, turkeys and even guinea-fowl. Of all of them, geese are the best grazers, then turkeys, ducks, chickens and finally guinea-fowl.

Egg Production

Free-range egg production is really the only system that can be considered as compatible with organic principles. It seems that the recent tremendous interest and growth in free-range production has now reached a level where it is beginning to meet the demand at the premium prices currently paid. This means that there is not a great deal of room for expansion in the conventional market. The organic market, on the other hand, is still virtually untapped, if you can get supplies of organic feed.

The question of which breed to choose is a complicated one. The standard brown hybrids such as the Warren and the Hi-sex tend to be too 'soft feathered' for the outdoor life. They certainly produce the most eggs with the least feed, but they do not range very well and some may hardly go out at all. The Shaver is better but the best is the Arbor Acre. Bred for free range, it is heavier, much more adventurous and will forage better than the others. Steer clear of the pure breeds for commercial production. The younger you buy in your birds the better, for then you can start them on an organic diet all the sooner. Failing that, try to obtain pullets that have been reared on free range so that they do not spend their first months with you having to find out what life is really all about. Their health and vigour should repay any extra cost that might be involved.

The EEC has now laid down a definition for free range which allows a maximum density of 1000 birds to the hectare. This is quite high for permanent stocking and organic standards now only allow 625 birds to the hectare. The pasture should be divided into at least two, so that you can rest one part whilst the other is in use. During winter the birds can range over the whole lot. The land should also be subject to a normal rotation so that only one batch uses the pasture before it goes back into other crops. If it is permanent pasture, you should alternate each batch of hens with a year of other stock to keep the parasites down.

The new free-range units that have sprung up to meet the recent demand have several thousand birds in one house. It is difficult to accept this as being in keeping with the principles of free range, particularly as some birds may hardly venture out at all. Separate small groups of between 150 and 200 birds make for a more natural situation, if more costly and more labour-intensive. Houses to accommodate these will have the advantage of being mobile.

Feed laying birds freely. Any restriction will affect their egg laying before anything else. The feed should be about 16 per cent protein and can be made up predominantly of wheat, then oats. Maize and barley can also be included. The protein can be supplied by 16 per cent soya and fish meal in the proportion of 4 to 1, or somewhat less meat and bone meal. Include also seaweed powder at 1 per cent and limestone flour at 5 per cent, and provide grit in a separate hopper. The hens will benefit from a covered scratching area where some whole grain cereals can be scattered every day. This will help to protect the rest of the field from being scratched about too much.

Lighting is allowed up to twelve hours a day, according to the Standards. This is well below the conventional amount and may not have much impact. However, it is less important on free range where the seasons are much more immediate to the birds. Lighting has the most effect when it is controlled to give an increasing day length, if this can be worked into the limited hours that are allowed. Two ways to compensate, if you do not use lighting, are to give an extra one and a half per cent of protein in the feed during the colder half of the year, and to ensure that the birds come in to lay about October or November. They will then have the naturally increasing day length when they most need it.

There is great debate about advantages and drawbacks of moulting and allowing a second year of production. Unfortunately the cost of feeding during the unproductive time of the moult is about equal to the cost of buying new birds. Those in their second season will lay probably a quarter less eggs of slightly larger size. The moulting process is also a critical time for the hens, although keeping them for just one laying season does not seem very good organic practice. That aside, your choice will rather depend on two factors: first, whether you have a good market for the small eggs that pullets produce initially and which usually go for processing at very low prices; and second, whether you have a good market for the culls which also go for very low prices due to the high numbers available from conventional producers.

Table Bird Production

Organic table bird production is even more of a good prospect than organic eggs. Again, the organic principles mean that they must be free range, but free-range table birds are few and far between. The demand is still fairly embryonic, but is surely set to take off just as free-range eggs have. Turnover of birds is obviously considerably higher than with layers, which might suggest that the system can be even more versatile. However, your outlets will undoubtedly require a steady flow of birds and this will only be achieved by establishing a constant and regular succession of batches coming on all the time.

There are several standard hybrid varieties for table production, and all of them are much better than the pure breeds. Their feed conversion ratios and growth rates are phenomenal. Probably the best known is the Ross and it is widely available. Conventional broiler houses produce 1.5 to 2kg birds in about eight to ten weeks. This high speed turnover is helped by the smaller sizes that are now preferred for frozen birds. The fresh market, on the other hand, which will be more the target for organic producers, can take something larger, thus allowing a slightly less hectic schedule.

The best way to start is to buy day-old chicks. These will need to go straight into a brooder for the first four to six weeks, depending on the weather. The alternative is to buy them when they are off heat and ready to go out. It does not take much of a mathematician to work out that from coming off heat to slaughter can be a very short period, and does not leave much time out on free range. Nevertheless, those few weeks, actually at least half the bird's life, make all the difference, as anyone who has tasted one will testify.

Feeding should again be ad lib. Whilst in the brooder, feed something in the region of a 22 per cent protein mix. This can be provided by about equal quantities of fish meal, soya meal and meat and bone meal, making up the 20 per cent inorganic portion. The best cereals are wheat, then maize, then barley, all ground up fairly fine. Include seaweed powder but not limestone flour and allow access to grit. Once the chicks are established outside, the feed can be changed to something nearer 19 per cent protein. The meal can be coarser, with equal quantities of wheat and barley, and oats instead of maize. Exclude fish meal from the protein, but include limestone flour at 1 per cent. In all these mixes, but not the chick rations, pea or bean meal can be used to substitute an equal weight of half soya and half cereal.

With a succession of batches at probably two or three week intervals, your paddocks must be well worked out and relatively easy to move or change about, both to accommodate different grazing requirements at different times of the year and to allow some rotation of grazing animals. In other words the fencing must not be permanent, except perhaps the perimeter fence, and should be relatively easy to move. This is now easily done with the new electric netting for chickens. It is based on the same principle as electric netting for sheep, and should keep out foxes as well as keep in the chickens. Only a mains fencing unit will produce sufficient power to overcome any short-circuiting that will frequently occur from the vegetation.

Chickens can be beset by many different pests and diseases, but because of their low individual value it is seldom worth treating individual cases. The situation obviously changes if many birds start showing symptoms. However, good husbandry in the form of rotation of pastures, moving of houses, providing adequate space at feeders and drinkers and sufficient nest boxes for laying birds will help to lessen the risks. The other very important aspect is careful attention to good hygiene, of which the main points are: always creosote the houses between batches, or at least once a year for table birds, and keep the houses well littered; do not mix batches of birds; place feeders and drinkers so that they cannot be soiled and renew the water every day; keep rodents out; and isolate any birds that show signs of ill health.

The most common problem is likely to be worms, but these will only get out of control if the conditions are too intensive and too dirty. A clove of garlic in the drinker, renewed every month, will help prevent them, as will wild garlic in the pasture. For treatment, use garlic and male fern root chopped in the feed, rue chopped in the feed or as a tea added to the water and eucalyptus oil in the water. Move to new pasture. Coccidiosis is usually prevented by the constant inclusion of low levels of drugs in the diet. This is allowed for growing birds but not for adults. Isolate, fast for a day with only water, then feed a mildly laxative diet (including bran and lots of green food) with a chopped clove of garlic a day for ten days. Move to new pasture and lime the old one.

Skin parasites are often the constant companions of poultry. Creosote the houses, particularly the crevices and cracks around the roosts and nest boxes, to kill the red mites. Add derris to the dust baths to kill lice. Do this every few days for about a month to ensure you catch all birds and get the next generation of lice as well. If it is really bad, spray the insides of the houses as well, at the beginning, middle and end of that month.

TURKEYS

People cry out for naturally reared turkeys, as their commercial production becomes ever more intensive. The main demand is obviously at Christmas, but a steady trickle is needed throughout the year. The one problem is that turkeys and chickens do not mix, for the former are very susceptible to the disease blackhead whilst the latter are fairly resistant carriers. Preventive treatment is allowed for young birds, but the safest remedy is not to keep both types on one farm and to be fastidious about keeping their litter clean and, therefore, topped up. Try to obtain a nosode for it, and for salmonella to which they are also susceptible. Most other ailments are similar to chickens and can be treated the same, remembering particularly that prevention is always better than cure.

Turkeys tend to require more cosseting than chickens when very small, including higher brood temperatures and better draught exclusion. After that, the reverse is almost true and they need plenty of fresh air. They also tend to be better grazers than chickens. If you buy them off heat, which is probably the best course of action, at least until you gain experience with them, wean them gently off their commercial feed over the next few weeks and on to the chicken grower rations, which should include grit and limestone flour. Barley and oats can be supplied in another hopper to make up to half of their feed. The design of the hoppers is quite important, for turkeys will not clear up spilt food and they do not like having their heads restricted.

Housing design is also important. They need fresh air to the extent that one side of their house should be just wire netting. Alternatively, as they get older, you may be able to teach them to roost higher up out of harm's way, in, or even on top of a barn or in trees. It is normal practice to separate the sexes at about ten to twelve weeks, but this should be less necessary on free range.

GEESE

Here again the main demand is at Christmas and very good prices are paid for them. Recently there has been a small revival of interest in the Christmas goose, probably because of the increasing tastelessness of commercial turkeys. Geese are not suited to intensive production. They are excellent grazers, but are not very efficient and tend to consume vast quantities of grass, particularly just when it stops growing as the winter draws on. However, the adults can survive on grass alone if they have access to enough, perhaps supplemented in the winter months and before they start laying.

Geese lay early and are not very prolific. If you want to produce your own you will get more eggs by collecting them from your breeding trios and using an incubator. But goose eggs are notoriously difficult to hatch out artificially, so you may be better advised to buy in the goslings and concentrate on fattening them. You might then be able to work out a more continuous programme. Goslings only need heat for the first two weeks. Reduce it progressively over the next week and provide them with reasonable cover for the next two. Thereafter, basic shelter is sufficient, provided there is fox protection.

The stocking density should not go above 125 geese per hectare, and, as with all poultry, the grazing quality must be good with a fine, close sward. Feed the standard high-protein starter ration whilst on heat, phasing it out gradually over the hardening off period and then feed only grain (mainly barley) and grit. Restrict this to about half a kilogram per bird, per week, until about six weeks before killing, when it should be given ad lib. You can bring them in for the last ten days to stop the fat becoming too yellow.

DUCKS

Some breeds of ducks are far more prolific egg producers than chickens, but somehow they have never caught on commercially. Instead, ducklings for the table are very popular, although you may have to work quite hard to establish a viable organic market, as most go to the catering trade which is fairly well supplied. The Aylesbury is the main breed and it is probably best to buy in day-old ducklings in a regular succession.

Feeding, brooding and housing can be much the same as for geese, changing over to the outside and grain feeding at the same time too. They eat grass fairly well and also slugs and snails, possibly making them good companions for the smaller organic grower who is troubled by these pests. However, they need to fatten and be ready at ten weeks old, when they moult and get set back somewhat, so the corn should be fed ad lib. Both geese and ducks are hardy birds that do not easily fall prey to disease, but both are quite highly strung and need gentle and quiet handling.

OTHER LIVESTOCK

If you want to try something different, or you do not feel that your conditions suit any of the more usual livestock, there are others you could consider. Up in the hills, deer farming is becoming popular. It has not yet become intensified, so needs lots of space, high fencing and careful attention to marketing.

If you are fortunate enough to have a good supply of clean running water on the farm, then fish farming might be a possibility. This has become an intensive business in recent years, with diseases building up and such artificial practices as the inclusion of colourants in the feed to ensure an attractive flesh colour. Surely there is scope for an organic fish farm to bring back to farmed trout the taste of their wild counterparts. It is a pity that carp are not considered edible in this country, for they would fit very neatly into the nutrient cycle of an organic farm.

On the poultry side, guinea fowl might be worth considering. They are hardy, generally free of disease, and very much at the top end of the quality range. But they are likely to be more of a sideline than a main enterprise. They are insect eaters rather than grazers, which might be very beneficial for the farm as a whole, and the adults will hardly need any housing, as they prefer to roost in trees. Perhaps their two biggest drawbacks are that they require good, high fencing to keep them in, and they are rather noisy.

Another very beneficial type of livestock is that wonderful creature, the honey bee. Unfortunately it is almost impossible to produce organic honey in Britain, despite its availability, as bees forage up to several miles and will not therefore confine themselves to organic crops. You would have to live in the middle of unfarmed country to be sure that your bees were organic. Then there is the problem of sugar feeding in the winter, and the fact that their labour requirements clash with other farming activities throughout the spring and summer. All in all, bees make an excellent addition to the balance of an organic farm, but they are unlikely to be a major organic enterprise.

CONCLUSION

Ending on a note of balance rather than of economics sums up organic agriculture quite well. It is not for the faint-hearted nor for those who want to make a fast buck; neither approach is likely to succeed, as the organic approach requires a depth of understanding that can only come from a profound commitment to this type of farming. For those who share this conviction, the satisfaction and fulfilment that they will gain from living out their ideals will easily be rewarding enough to outweigh the harder work and possibly lower returns that are the organic farmer's lot. Good luck!

Appendix

CONVERSIONS

1 centimetre   0.39 inch
1 metre   1.09 yards

1 litre   0.22 gallons

1 kilogram   2.20 pounds
1 tonne   0.98 ton

1 hectare   2.47 acres

Useful Addresses

Agricultural Training Board
32–34 Beckenham Road
Beckenham, Kent BR3 4PB
01 650 4890

Joseph Bentley Ltd
(organic fertilisers and minerals)
Barrow on Humber
South Humberside DN19 7AQ
0469 30501

Biodynamic Agricultural
Association
Woodman Lane
Clent
Stourbridge
W. Midlands DY9 9PX
0562 884933

Booker Seeds Ltd
(agricultural and horticultural
seeds)
Boston Road
Sleaford
Lincs NG34 7HA
0529 304511

British Organic Farmers
86 Colston Street
Bristol, Avon BS1 5BB
0272 299666

Chase Organics Ltd
(seeds, seaweed, sundries)
Terminal House
Shepperton
Middlesex TW17 8AS
0932 221212

Elm Farm Research Centre
(soil analysis service)
Hamstead Marshall
Nr Newbury
Berks RG15 0HR
0488 58298

Elsoms Seeds Ltd
(agricultural and horticultural
seeds)
Pinchbeck Road
Spalding
Lincs PE11 1QG
0775 5011

Emerson College
(school of biodynamic agriculture)
Pixton
Forest Row
Sussex RH18 5JX
034 282 2238

Fertosan Products Ltd
(herbal slug destroyer, among
other products)
2 Holborn Square
Lower Tranmere
Birkenhead
Merseyside L41 9HQ
051 647 5041

Free Range Egg Association
37 Tanza Road
London NW3 2UA
01 435 2596

Glasshouse Crops Research Institute
Worthing Road
Littlehampton
W. Sussex BN17 6LP
0903 716123

Guild of Conservation Food Producers
PO Box 157
Bedford MK42 9BY
0234 61626

Henry Doubleday Research Association
National Centre for Organic Gardening
Ryton on Dunsmore
Coventry CV8 3LG
0203 303517

Humber Fertilisers PLC
(one organic manure in their list of semi-organics)
PO Box 27
Stoneferry
Hull HU8 8DQ
0482 20458

International Federation of Organic Agriculture Movements
161 Dom. des Bois Murés
06130 Grasse
France

International Institute of Biological Husbandry
Abacus House
Station Approach
Needham Market
Ipswich, Suffolk
0449 720838

Koppert UK Ltd
(biological control, insecticidal soap)
PO Box 43
Tunbridge Wells
Kent TN2 5BY
0892 36807

McCarrison Society
36 Bowness Avenue
Headington
Oxford OX3 0AL
0865 61272

Microbial Resources Ltd
(Biological control products)
Theale Technology Centre
Theale
Berks RG7 4JW
0734 303707

National Institute of Agricultural Botany
Huntingdon Road
Cambridge CB3 0LE
0223 276381

National Vegetable Research Station
Wellesbourne
Warwick CV35 9EF
0789 840382

OGA Packaging
c/o John Blake
Old Bridge
South Petherton
Somerset TA13 5LR
0460 40855

Organic Advisory Service
c/o Mark Measures
Elm Farm Research Centre
Hamstead Marshall
Nr Newbury
Berks RG15 0HR
0488 58298

Organic Farmers and Growers Ltd
(organic fertilisers, machinery)
9 Station Approach
Needham Market
Stowmarket
Suffolk IP6 8AT
0449 720838

Organic Growers Association
86 Colston Street
Bristol BS1 5BB
0272 299800

George A Palmer Ltd
(one organic manure in their list of semi-organics)
Oxney Road
Peterborough PE1 5YZ
0733 61222

Rhone-Poulenc (UK) Ltd
(Redzlaag)
Hulton House
161/166 Fleet Street
London EC4A 2DP

Sea Trident Ltd
(seaweed products)
Sarum House
Oak Park
Dawlish
Devon EX7 0DE
0626 862489

Small Farmers Association
PO Box 71
Beaufort House
15 St Botolph Street
London EC3A 7HR
01 247 4187

Soil Association
86 Colston Street
Bristol BS1 5BB
0272 290661

Soil Fertility Dunns Ltd
(Gafsa rock phosphate)
Hartham
Corsham
Wilts SN13 0QA
0249 712051

A L Tozer Ltd
(horticultural seeds)
Pyports
Cobham
Surrey KT11 3EH
0932 62059

Working Weekends On Organic Farms
19 Bradford Road
Lewes
E Sussex BN7 1RB

Further Reading

BOOKS

Balfour, E. B. *The Living Soil and the Haughley Experiment* (Faber and Faber, 1975)
Hart, E. *Sheep, a Guide to Management* (The Crowood Press, 1985)
Howard, A. *An Agricultural Testament* (Oxford University Press, 1943)
Koepf, H. H., Pettersson, B. D., Schaumann, W. *Biodynamic Agriculture* (Anthroposophic Press, 1976)
Levy, J. de B. *The Complete Herbal Handbook for Farm and Stable* (Faber, 1984)
McCarrison, R., Sinclair, H. M. *Nutrition and Health* (Faber and Faber, 1961)
Mackenzie, D. *Goat Husbandry* (Faber and Faber, 1972)
Pfirter, (*et al.*) *Composting* (Co-operative Migros Aargau/Solothurn, 1981)
Thear, K. *The Complete Book of Raising Livestock and Poultry* (Martin Dunitz, 1981)
Thear, K. *Practical Chicken Keeping* (Ward Lock, 1983)
Woodward, L. *Green Manures* (Elm Farm Research Centre, 1982)

OTHER PUBLICATIONS

Grower (weekly), 50 Doughty Street, London WC1.
Home Farm (bi-monthly), Broad Leys Publishing Co, Widdington, Saffron Walden, Essex.
New Farmer and Grower (quarterly), OGA and BOF, 86 Colston Street, Bristol.
Review (quarterly), Soil Association, 86 Colston Street, Bristol.

Index